NASA'S BEES

Rob Waugh is a leading international science and technology journalist, having written about gadgets, apps and business technology for dozens of newspapers, magazines and websites over the last 25 years, including the New York Post, Daily Mail USA, the Telegraph, the Daily Mail, the Guardian and many more. He has a particular interest in artificial intelligence, and has attempted to tame many domestic robots over the years, with varying degrees of success.

This edition published by Shelter Harbor Press by arrangement with Elwin Street Productions Limited

Copyright © Elwin Street Productions Limited 2023
Conceived and produced by
Elwin Street Productions Limited
10 Elwin Street
London E2 7BU
United Kingdom

Additional text by Sam Hartburn
Illustrations by Jason Anscomb, Rawshock design.
Photo credits: Shutterstock, except page 169: craiyon.com.

All rights reserved. No part of this publication may be reproduced, stored in a retrieval system, or transmitted in any form or by any means (including electronic, mechanical, photocopying, recording, or otherwise) without prior written permission from the publisher.

SHELTER HARBOR PRESS
603 W. 115th Street, Suite 163
New York, NY 10025
www.shelterharborpress.com

For sales in the US and Canada, please contact info@shelterharborpress.com

ISBN 978-1-62795-190-6

10 9 8 7 6 5 4 3 2 1

Printed in China

NASA'S BEES

FIFTY EXPERIMENTS
THAT REVOLUTIONIZED
ROBOTICS & AI

ROB WAUGH

Shelter Harbor Press
New York

Contents

INTRODUCTION		**8**
1 DREAMING OF ROBOTS: 322 BCE-CE 1700		**10**
322 BCE	**When did we first start dreaming of robots?**—Aristotle	12
	Aristotle's early optimism	
400-300 BCE	**What was the first working automaton?**—Hero of Alexandria	14
	How Hero of Alexandria's automata came to life	
100 BCE	**Can machines predict our future?**—Unknown	17
	How the Antikythera mechanism computed the planets	
800-873	**Can robots make melodies?**—Banu Musa	20
	How ninth-century Baghdad led to computerized music	
1200-1300	**Can our thinking be mechanized?**—Llul	23
	How Ramon Llull's volvelle automated thought	
1495	**Fancy drawings or feasible science?**—da Vinci	26
	Da Vinci's experiments with automation	
1600s	**How do karakuri dolls work?**—Omi	29
	How karakuri puppets taught Japan to love robots	
2 INDUSTRY AND AUTOMATION: 1701-1899		**32**
1701	**How can sowing be made more efficient?**—Tull	34
	How Jethro Tull's seed drill broke new ground	
1763	**What will happen next?**—Bayes	37
	How Bayes' theorem lets us predict the future	
1804	**Can a machine take orders?**—Charles	40
	The dawn of programmable machines	

1832	**How did mathematics find its engine?**—Babbage and Lovelace	43
	Babbage, Lovelace and the computing machine	
1871	**How did mechanization change publishing?**—Hoe	46
	Hoe's Lightning Press	
1898	**Who invented remote-controlled vehicles?**—Tesla	49
	Nikola Tesla's "teleautomaton"	

3 THE DAWN OF MODERN ROBOTICS: 1900–1939 — 52

1914	**Computer vs. man**—Torres-Quevedo	54
	The first (unbeatable) chess-playing automaton	
1914	**What does "robot" mean?**—Čapek	58
	How Karel Čapek's play invented the word "robot"	
1925	**Can a robot drive itself?**—Houdina	60
	How the Houdina "American Wonder" sparked self-driving cars	
1927	**Can a robot respond to instructions?**—Wensley	63
	How Herbert Televox did a human job	
1928	**What should "man machines" look like?**—Lang	66
	From movies to reality	
1938	**What use was Pollard's patent?**—Pollard	69
	Why a "position-controlling apparatus" paved the way for robot arms	

4 DEVELOPING INTELLIGENCE: 1940–1969 — 72

1942	**Are robots above the law?**—Asimov	74
	How Asimov's "Laws of Robotics" help us imagine a human–robot society	
1944	**How did women help ENIAC?**—Mauchly and Holberton	77
	The hard-working thinking machine	
1949	**Can machines think like us?**—Berkeley	80
	How "giant brains" helped us imagine a computer in every home	
1950	**How does a machine pass the Turing test?**—Turing	83
	Assessing a machine's ability to show intelligent behaviour	

1951	**What is SNARC?**—Minsky	86
	The first neural network machine that "learned" like a human brain	
1956	**When was Artificial Intelligence born?**—McCarthy	89
	The Dartmouth Conference	
1960	**Can a machine look after itself?**—Chubbuck	92
	How the Beast learned to feed itself	
1961	**Can a robot do a human job?**—Devol	95
	How robots revolutionized manufacturing	

5 SURVIVAL OF THE FITTEST: 1970–1998 — 98

1970	**How did Shakey think?**—Rosen	100
	Why Shakey's navigation changed the world	
1987	**Can robotics be used to treat cancer?**—Adler	104
	CyberKnife radiosurgery	
1990	**Can machines learn from their behaviour?**—Mataric	107
	How Toto helped machines to "learn"	
1990s	**Can robots express emotion?**—Breazeal	110
	Kismet and social intelligence	
1993	**Can robots swim under water?**—Triantafyllou	113
	How Robotuna helps us explore the seas	
1997	**Who is better at football?**—Kitano et. al	116
	The goal of the Robocup	
1997	**How did a computer win at chess?**—Hsu and Campbell	119
	What Deep Blue taught us about intelligence	

6 ROBOTS AT HOME: 1999–2011 — 122

1999	**Can robots replace our pets?**—Doi and Fujita	124
	Why people loved Aibo	
2000	**Can a robot stand on its own two feet?**—Shigemi	127
	How ASIMO played soccer with a president	

2001	**Can robots kill?**—General Atomics How the MQ-9 Reaper changed warfare	130
2001	**Why are slugs scared of robots?**—Kelly, Holland and Melhuish A slimy diet for autonomous robots	133
2002	**Can a robot do our chores?**—Angle, Greiner and Brooks How efficient is a robot cleaner	136
2003	**How far can robots go?**—Squyres The Mars rover Opportunity	139
2005	**How do cars drive themselves?**—Thrun How the DARPA Grand Challenge created self-driving cars	142
2011	**Can robots help us walk?**—Sankai The life-changing HAL exoskeleton	145

7 SCI-FI BECOMES REALITY: 2011–PRESENT   148

2011	**Can humanoids help astronauts?**—Badger What Robonaut 2 taught us	150
2015	**Can a robot be a police officer?**—Stephens The pros and pitfalls of the knightscope security robot	152
2016	**How did a computer learn to win at Go?**—Hassabis From Alphago to Muzero	155
2016	**Can robots become radicalized?**—Lee Why chatbot Tay only lived for a day	158
2016	**How did Sophia get her citizenship?**—Hanson The robot who was granted citizenship of Saudi Arabia	160
2018	**Can a machine be curious?**—Gannon How Mimus can help us coexist with AI	163
2019	**Can a bee fly in space?**—Bualat How astrobees might help us reach Mars	166
2022	**Will AI take over the world?**—Altman How ChatGPT disrupted tech overnight	169

Index	172
Glossary	174
Sources	175

Introduction

The past year has seen a frenzy erupt around artificial intelligence, thanks to 'generative AI' apps such as Stable Diffusion, Dall-E and ChatGPT, and their eerily human-like way of generating text and images. With billions in investment pouring into the technology, AI is set to reshape everything from the way we interact with computers to the jobs we do.

AI and robots are already all around us. When you speak to your phone's voice assistant, or use a "smart" thermostat, you are relying on AI. The products we buy are sorted with machine efficiency in warehouses staffed by robots.

In everything from space exploration to surgery, robots are set to take over jobs too risky and difficult for human beings (and expand the reach of human experts across continents and even into space). When the first human boots step onto the surface of Mars, some of the crew won't be human: they will be robots.

So how did we get here? This book examines 50 milestones in robotics and artificial intelligence, from the first moment ancient people imagined artificial servants through to cutting-edge machines set to shape humanity's future.

Perhaps the biggest surprise is how long the human race has been obsessed with automatons and machines that think. In the fourth century BCE, ancient Greek philosopher Aristotle imagined a future where automated tools took over mundane tasks from human beings. Ancient Greek scientists, meanwhile, designed everything from vending machines to animated "maids" which poured wine, powered by pneumatics, gears and cogs.

In ancient China, a text describes a strange, animated man shown off by an inventor to a baffled king. In ninth-century Baghdad, three brothers created a book of strange devices and automata, including a flute-player powered by flowing water which played pre-programmed tunes.

Hundreds of years before there would be machines that could process ideas, thirteenth-century mystic Ramon Llul designed a machine of rotating paper discs to convert people to Christianity. He is now seen as a "prophet" of computer science. In an even more visionary leap, fifteenth-century polymath Leonardo da Vinci designed a mechanical knight which waved its arms – and a "self-driving" cart which may have been programmable.

In the Industrial Revolution, machines such as the Jacquard loom set the stage for the modern world: the punch cards used by the device inspired computing pioneer Charles Babbage, and paved the way for twentieth-century computing. By the 1960s, America used 500 billion punch cards per year.

Even as robotics and artificial intelligence gathered pace in recent centuries, many pioneers have gone relatively unsung, from Nikola Tesla's demonstration of a remote-control boat in 1898 (which was considered so unearthly some observers thought there was a monkey inside the machine) to the first driverless cars (or "phantom autos") which took to the streets in the 1920s, sending pedestrians leaping for their lives. Alan Turing's work during World War II on breaking the Enigma cipher is now well known, but across the channel, another computer pioneer labored on a machine that would be destroyed by Allied bombs and remain unknown to the wider world until after the fall of Berlin.

The latest breakthroughs in robotics and AI offer us a glimpse of our own future. Nasa's Astrobees are floating cubes which "fly" through the microgravity of the International Space Stations using air nozzles.

When Google's AlphaGo defeated the best human player at the ancient board game Go in 2016, the team behind it moved on to developing systems which can beat games without ever being told the rules. It could lead to an artificial intelligence system which can solve real-world problems on its own, without being told what to do, or how to do it. We live in a world where science fiction is becoming reality: this book tells the story of how it happened.

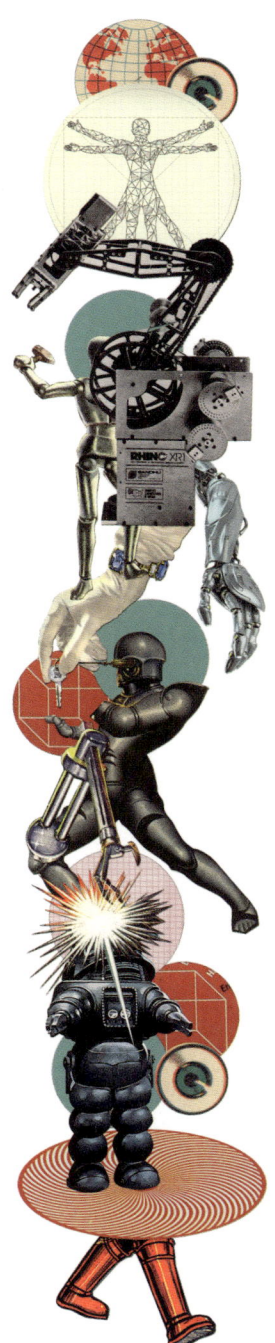

CHAPTER 1: Dreaming of Robots
322 BCE–CE 1700

Long before there were technologies that could deliver "living" beings made of metal, ancient people dreamed of "automata," with Greek myths describing huge bronze men, and objects that sprang to life, animated by the magic of the gods.

Philosophers such as Aristotle imagined a world where living tools would get rid of slavery forever. Technologies such as pneumatics began to lay the foundations for animated machines, with automata that chirped like birds and poured wine.

   Before the first century CE, inventors created steam engines and vending machines—and theaters filled with automata, using special effects to bring the gods to life.
   But it wasn't just in ancient Greece where lifeless objects were coming alive, powered by water, steam, and air: in Baghdad, three brothers created the first programmable device on Earth, while, in ancient China, a tale told of a strange, animated puppet that could walk and talk.

322 BCE

RESEARCHER:
Aristotle

SUBJECT AREA:
Automated servants

CONCLUSION:
Imagined automated tools that could take the place of workers

WHEN DID WE FIRST START DREAMING OF ROBOTS?

ARISTOTLE'S EARLY OPTIMISM

The word "automaton" comes from Homer's *Iliad*, the ancient Greek epic poem set during the Trojan War, thought to have been written in the eighth century BCE. In the poem, Hephaestus, the blacksmith of the gods, has several marvelous machines that work for him, including bellows that spring to life, various humanlike servants made of gold and silver, and a set of gold and silver guard dogs. But perhaps the most intriguingly robotlike of these servants are Hephaestus's animated tripods, which are described as "automatons."

Greek mythology did feature other "artificial men." For instance, the giant Talos, also crafted by Hephaestus, a gigantic metal guardian who meets his end when Jason and the Argonauts remove a huge nail from his heel, allowing all the "ichor"—the bloodlike fluid that courses through the gods veins—to run out of his bronze body. Talos was immortalized in animated form by the stop-motion animation of Ray Harryhausen in the Hollywood classic *Jason and the Argonauts* in 1963. But it was a few hundred years after Homer's *Iliad* when Aristotle considered the idea of automated tools that could take the place of slaves, and how such machines might fit into society.

A philosopher and scientist born around 384 BCE in Greece, Aristotle was a student of the philosopher Plato, and went on to be the tutor of Alexander the Great.

Liberation through technology

Slavery was a fact of life in ancient Greece, with well-to-do families owning at least one slave. So it was in the terms of a world filled with slaves that Aristotle imagined his automated tools. Aristotle wrote, "If each instrument could do its own work, at the word of command or by intelligent anticipation, like the statues of Daedalus or the tripods made by Hephaestus [. . .] a shuttle would then weave itself, and a plectrum would do its own harp playing. In this situation managers would not need subordinates and masters would not need slaves."

There are two ways to read the passage: one is that Aristotle is describing an absurd situation, mocking the idea that society could ever be upended in this way. In the other, he's describing something that he believes could happen, and that he's hopeful that the technology will one day arrive to liberate workers and slaves.

Either way, Aristotle's idea that automation would mean a day of liberation for slaves was optimistic. In the Industrial Revolution, for instance, cotton was often used in the first factories, as machines could process cotton fibers better than wool. But the cotton was still picked by slaves on plantations in America.

More than a thousand years after Aristotle, his ideas of machines that would take the place of human workers began to become reality, in the form of devices such as the Jacquard loom, which in the nineteenth century vastly sped up the weaving process (see page 40).

400–300 BCE

RESEARCHER:
Hero of Alexandria

SUBJECT AREA:
Automata

CONCLUSION:
Hero created animated characters, a theater, and even a steam engine

WHAT WAS THE FIRST WORKING AUTOMATON?

HOW HERO OF ALEXANDRIA'S AUTOMATA CAME TO LIFE

The idea of "automata" fascinated the ancient Greeks, with their myths of bronze men crafted in a blacksmith's forge. As technologies such as hydraulics, waterpower, and steam power emerged, Greek scientists and writers gained the power to create their own "living" creatures of metal and wood, ranging from animated animals to a real-life steam engine.

The machines were built using technological trickery using gears and ropes, and were often toys, designed to entertain, or to create a magical effect.

By the second century BCE, Philo of Byzantium described pneumatic devices including a human-shaped "maid" that automatically dispensed wine and water into a cup (ancient Greeks and Romans often mixed wine and water before drinking).

But perhaps the most prolific creator of automata was Hero of Alexandria, who died around 70 CE. Hero was a highly influential mathematician and geometer (of whose several works have survived to the present day), but also wrote designs for a huge number of strange toys and animated creatures.

Chirping Birds

The machines themselves have not survived, but Hero's designs and descriptions are clearly practical and workable: one design sees a group of mechanical birds that start to sing together, then fall silent when a metal owl turns around to look at them. The hydraulic automaton, adapted from an earlier design of Philo's, is driven by hidden tubes and siphons filled with water, which revolve the tubes that the birds stand on. The chirping is created by forcing air through containers with water in the bottom.

Hero described a huge number of such machines, including a theater that could stage plays "acted" by automatons, powered by weights, cogs, and sand running into a container, as in an hourglass. The theater was a firework display of different special effects, driven by ropes attached to the automata characters. A play saw fire being lit on an altar in front of the god Dionysus, then milk running from his staff as he spilled wine over a panther. His followers dance to the sound of drums. The automata were pulled with cords wrapped around drums, with different lengths of rope allowing different "characters" to move at different times.

Hero wrote of Athena, a character in one play, "One cord will raise her by pulling from behind her hips, and will keep her balanced. After this cord is released, another, set around her waist, pulls her in a circle, until she returns to the spot from which she started."

Air and Steam

The wine and milk are poured using pneumatic techniques, something Hero explored in another book, *Pneumatica*, where running water and air pressure are used to create hooting owls and animated mythical heroes, many of whom interact with their audience.

Hero writes, "On a pedestal is placed a small tree, around which a serpent or dragon is coiled; a figure of Hercules stands near, shooting from a bow and an apple lies on the pedestal: if anyone raises, with the hand, the apple a little from the pedestal, Hercules shall discharge his arrow at the serpent and the serpent hiss."

Many of Hero's inventions were built to be used in temple settings, to provide "magical" effects, including a device that automatically opened temple doors when a fire was lit, and a vending machine that dispensed water in response to a five-drachma coin being inserted.

Perhaps Hero's most ahead-of-its-time invention (and a famous case of missing the technological point entirely) was a rotating ball that was driven by steam

power. Hero wrote, "Place a cauldron over a fire, a ball shall revolve on a pivot."

It would be more than a millennium and a half before steam power would revolutionize industry in Europe and around the world, paving the way for inventions such as the rotary printing press (see page 46).

"That, Sire, is my own handiwork"

But the technologies used by inventors such as Hero may not have been unique. One Chinese text suggests that some form of automata may have been shown off by an inventor in a Chinese royal court—and it may even predate Hero's inventions.

The Chinese Lie Zi text includes a strange account of an encounter between a king and an animated man in the fourth century BCE: "'Who is that man accompanying you?' asked the King. 'That, Sire, is my own handiwork. He can sing and he can act.' The King stared at the figure in astonishment. It walked with rapid strides, moving its head up and down, so that anyone would have taken it for a live human being. The artificer touched its chin, and it began singing, perfectly in tune. He touched its hand, and it started posturing, keeping perfect time."

The tale clearly has fictional elements (the King gets annoyed with the automaton's creator, who dismantles it in front of him, revealing how the automaton loses its senses and abilities one by one as its master removes its internal organs), but it raises intriguing questions about what real automata may have existed in ancient China.

An account from the third century BCE describes a mechanical orchestra being made for the emperor, and by the Tang period (seventh to tenth century CE), automatons including an otter that caught fish and a monk begging were popular in the imperial court.

Both Chinese and Greek automata predated by more than a millennium the fashion for automata in eighteenth-century Europe and the karakuri puppets of Edo period Japan, but showcased simple versions of the technologies that would be used to animate ducks and demons in the centuries to come.

CAN MACHINES PREDICT OUR FUTURE?

HOW THE ANTIKYTHERA MECHANISM COMPUTED THE PLANETS

100 BCE
RESEARCHER:
Unknown
SUBJECT AREA:
Astronomical calculation
CONCLUSION:
The mechanism helped ancient Greeks predict eclipses and other events

In 1900, captain Dimitrios Kontos sent a team of sponge divers out to work off the coast of the Greek island Antikythera while waiting out a spring storm in their boat. Diver Ilias Stadiatis resurfaced with the arm of a bronze statue, saying that there was more down there.

He had found the wreck of a trading ship from the first century BCE. Among the treasures brought to the surface was a calcified lump filled with ancient gear wheels. Few ancient objects can match the mystery of the Antikythera mechanism, which, over the course of 120 years, has been slowly reassembled from that handful of pieces. Often described as the "world's first computer," the mysterious clockwork apparatus is only now close to being fully understood.

It took researchers some time to realize the magnitude of their discovery. The complexity of the gears inside the device was completely unknown in the ancient world, and remained so until the construction of the first cathedral clocks a thousand years later.

The Antikythera Mechanism Research Project, which brings together different research teams working on the mechanism, says that initially the machine was somewhat overlooked, due to fears that statements of its complexity were exaggerated by overenthusiastic researchers.

In fact, the opposite was true.

Like Clockwork

The discovery begged questions. Why has nothing similar ever been found? And how much could the device actually do? Researchers knew that the mechanism's functions related to astronomy, but it took decades to work out what those were.

The mechanism has fascinated researchers from classicists to astronomers to computer scientists, with several building replicas of what else might have been inside the incomplete mechanism in an attempt to understand how it worked.

The machine had the first known set of scientific dials, and thirty gear wheels, which were spotted on X-rays. It was fabricated from bronze sheet, and was covered in Greek inscriptions, which reveal that it was used as a sort of astronomical calendar.

It's believed that a central shaft (which is lost) would have turned a large main gear with each turn matching one solar year. There's a large dial that shows the position of the sun and moon and a ball for lunar phases. The machine would have allowed ancient people to predict astronomical events such as eclipses.

Sole Survivor

According to the Antikythera Mechanism Research Project, the reason that nothing like it has survived is fairly simple. Bronze was, at the time, not just highly valuable, but highly recyclable, and used for coinage. As a result, most surviving ancient bronze finds come from underwater locations such as shipwrecks, where there was no chance of the metal being melted down and reworked into something else.

The researchers also believe that there were probably other, similar machines: not only are other complex machines described in surviving Greek writings, but the mechanism shows no signs of the design having been modified as it was built, suggesting that the maker must have had experience of building similar machines.

The machine has inspired devotees to build their own replicas, including a working one built by Apple engineer Andy Carol entirely from Lego. Carol's machine doesn't look exactly like the mechanism (due to the limitations of Lego), but he believes its functions are very similar.

"It's an analog computer, which means it can't execute programs," says Carol. "The Antikythera mechanism, and my Lego version, are both just simple mechanical computers: you turn the crank at one speed and all the wheels move at another speed, which you've calibrated to have a particular meaning—in this case, predicting the cycles of astronomical bodies."

Rebuilding the Machine

Carol says that the analog computing power of the mechanism is similar to machinery used on battleships in World War II to compute distances. Others, including Michael Wright of London's Science Museum, have built their own partial replicas of the machine. In 2021, a team from University College London (UCL) reconstructed the gearing system that moved the front of the device for the first time. Previous X-ray research in 2005 had revealed how the mechanism could predict eclipses and calculate the motion of the moon.

But the UCL team used inscriptions unearthed by X-ray to reconstruct the cosmos display, with planets moving on rings, shown by marker beads. The team used an ancient Greek mathematical technique described by the philosopher Parmenides to work out how the maker of the Mechanism accurately represented the 462-year cycle for Venus and the 442-year cycle for Saturn.

Mechanical engineering Professor Tony Freeth said, "Ours is the first model that conforms to all the physical evidence and matches the descriptions in the scientific inscriptions engraved on the Mechanism itself. The sun, moon, and planets are displayed in an impressive tour de force of ancient Greek brilliance."

The team now hopes to rebuild the mechanism using the tools that would have been available to craftsmen at the time. But many, including the Antikythera Mechanism Research Project, believe that the machine may hold more mysteries yet to be discovered.

800–873

RESEARCHERS:
Jafar Muhammad, Ahmad, and al-Hasan ibn Musa ibn Shakir

SUBJECT AREA:
Automata

CONCLUSION:
Created a programmable flute player centuries ahead of its time

CAN ROBOTS MAKE MELODIES?

HOW NINTH-CENTURY BAGHDAD LED TO COMPUTERIZED MUSIC

In the ninth century CE, Baghdad was the richest city on Earth, with caliphs ruling over an empire bigger than the Roman Empire had been at its height. The city was also becoming the greatest center of science on the planet. Islamic scientists made breakthroughs in medicine, astronomy, chemistry, and mathematics (the word algebra comes from the Arabic "al-jabr").

Among the many advances found in the area at that time were automata that foreshadowed the abilities of robots many centuries later. Baghdad's Banu Musa (Musa brothers) created the world's first programmable device, a music player that could change tune and tempo—all many centuries before anything remotely comparable was invented anywhere else.

Court of the Caliph

The brothers were the sons of Musa bin Shakir, a robber turned astronomer-engineer. Like their father, the brothers were regulars of the Caliph's court (and not above political power-play against rivals). The brothers were named Jafar Muhammad, who specialized in geometry and astronomy, al-Hasan, who specialized in geometry, and Ahmad, who worked mostly with mechanics (and is thought to have been the driving force behind the brothers' automata). Often signing their works together, the Banu Musa brothers authored books on mathematics and astronomy.

They rose to prominence when the Caliph Al-Ma'mun became their guardian after their father's death. Al-Ma'mun founded the House of Wisdom (described as the most comprehensive library since the Library of Alexandria), and built astronomical observatories. The Caliph recruited the brothers to the House of Wisdom, and set them challenges including measuring a degree of latitude, which they did from the desert with an impressive degree of accuracy.

Science historian Jamal Al-Dabbagh wrote in the *Dictionary of Scientific Biography*, "The Banu Musa were among the first Arabic scientists to study the Greek mathematical works and to lay the foundation of the Arabic school of mathematics. They may be called disciples of Greek mathematics, yet they deviated from classical Greek mathematics in ways that were very important to the development of some mathematical concepts."

The Banu Musa made breakthroughs in measuring area and volume, in observing the sun and moon, and in measuring the length of a year. But despite authoring more than twenty works, including several that survive to this day, the brothers are probably most famous for their mechanical tricks.

Ingenious Devices

Their most famous work is the *Book of Ingenious Devices*, which describes a huge variety of "trick" jugs with amazing powers such as being able to pour two liquids in without mixing them, and pouring them out again separately (achieved by separate hidden compartments inside the jug).

Five hundred years later, Arab historian Ibn Khaldun wrote, "There exists a book on mechanics that mentions every astonishing remarkable and nice mechanical contrivance."

Most of the gadgets were designed as novelties, although among the 100 in the volume are some genuinely useful devices such as a clamshell grabber used for picking up objects underwater, and a bellows-type device to remove foul air from wells.

Another miraculous invention dispensed water in measured amounts (similar to how a sink works in a public toilet to conserve water). Some of the devices were variations on themes developed by ancient Greek writers; others were new.

Let the Music Play

But musical devices invented by the Banu Musa offered startling innovations, with one recognized as the first music sequencer (similar to those used by electronic artists today) and the first programmable machine.

Thought to have been built around 875 CE, the machine was designed so that it could continuously play tunes, powered by a steady current of water. It was constructed to look like a human flute player, playing a wind instrument with his fingers.

The brothers wrote, "We wish to explain how an instrument is made that plays by itself continuously in whatever melody, and also that we may change from melody to melody when we so desire."

A hidden chamber inside it used flowing water to provide air pressure to the flute instrument. This was not unique: previous Greek and Chinese automata had played wind instruments before. But those automata simply repeated the same pattern (or in many cases whistled due to air being forced through a pipe).

What put the brothers' creation on a different level was what was inside the figurine. Here was a pin-barrel mechanism similar to the ones later used in children's music boxes, which was propelled by the flowing water.

Importantly, the barrel could be changed and programmed, allowing the brothers to vary both the melody and the tempo of the music. It has been described as the first programmable device of any kind on Earth, as an ancestor to the first computers. In fact, experts have suggested that it was only in the twentieth century that comparable music devices arrived: sequencers, which play music from a series of instructions, just like the brothers' flute player's fingers. The flute-playing automata of ninth-century Baghdad has been described as an ancestor to the whole field of computer music.

CAN OUR THINKING BE MECHANIZED?

HOW RAMON LLUL'S VOLVELLE AUTOMATED THOUGHT

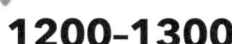

1200–1300
RESEARCHER:
Ramon Llul
SUBJECT AREA:
Automating thought
CONCLUSION:
The first way to think "mechanically," and a huge influence on later scientists

What does a thirteenth-century Christian mystic have in common with computer scientists in the twenty-first century? Not much, you might imagine. But Ramon Llul, a novelist and poet who was born in Majorca in 1232 and died in Tunis in 1315 (allegedly after being stoned by Muslims who he was hoping to convert to Christianity), is considered an inspiration by many computer scientists today.

At the age of thirty, supposedly while just at the moment of composing a bawdy love song, Llul experienced a mystical vision of Jesus on the Cross. It proved a turning point. Thereafter, he devoted himself to missionary work, traveling into North Africa and other areas in an attempt to convert local people to Christianity.

Llul is famous for promoting the Catalan language, and for his ideas on elections, which were hundreds of years ahead of their time. But oddly, the writing that endears him to modern computer scientists is actually a logical tool designed to convert Muslims to the Christian faith.

A Thinking Machine

Llul noted that attempts to convert Muslims to Christianity by public debate had not worked. He believed that to convert people, he had to find a mechanism that could prove and generate truths about God.

Llul used a machine called a *volvelle*, a revolving paper set of concentric circles, which he detailed in his book of philosophy *Ars Magna Generalis Ultima* (Great, General, and Ultimate Art). Volvelles were not unique to Llul, but the way he used the paper machine was.

For his "thinking machine," Llul is thought to have drawn inspiration from observing an astrological device called a *zairja*, which was used by Arab astrologers to generate ideas.

For Llul, the purpose of his logical machine was to break ideas down into units that can then be randomly connected to each other, producing all the possible combinations of arguments by spinning the concentric disks. On the outer disk were the nine names of God, while on inner disks were attributes of God.

Researcher Suzanne Karr, author of the journal article "Constructions Both Sacred and Profane," wrote, "When properly used, the triple layer of combinations of nine letters [...] answered questions about all creation and even the future, as well as inquiries intended to settle religious debates."

By spinning the machine, Llul would create random chains of association, which could reveal all aspects of the deity, automatically. On the volvelle, the names and aspects of God were shown as letters, and then the user could read back the combination of ideas that each position of the three disks (held in place by a central pin) revealed.

In previous versions of his idea, Llul had suggested using a treelike diagram from which readers could draw their ideas: but the revolving disks added an automated element, which influenced later thinkers.

Spinning ideas

The idea of using a machine to represent units of thought was revolutionary. Llul also created a volvelle called the Night Sphere, which used the position of stars to calculate time during the night (allowing a physician to administer medicine at the correct time). Such disks were used to calculate dates and astronomical events across Europe.

But Llul's idea of using the volvelle to connect ideas was to be a huge and direct influence on a man who built one of the very first ancestors of modern computers, seventeenth-century German inventor and polymath Gottfried Wilhelm Leibniz.

At just twenty, Leibniz published a dissertation called "On the Combinatorial Art," which suggested that human ideas could be broken down into units that could be represented by symbols (which he called "an alphabet of human thought").

He hoped to produce a logic-calculating machine ("the great instrument of reason"), which could answer any question and solve any debate.

Llul's Idea Come to Life

Leibniz described his idea as Llul's dream come to life, and it's this endorsement that has meant that Llul is seen by some as a founding father of modern computer science. "If controversies were to arise, there would be no more need for disputation between two philosophers than between two calculators," he wrote. "It would suffice for them to take their pencils in their hands and to sit down at the abacus, and say to each other (and if they so wish also to a friend called to help): 'Let us calculate!'"

Leibniz's famous rallying cry of "*Calculemus!*" ("Let us calculate!") offered an optimistic view of a future where machines could solve human problems. He hoped that such a machine could solve philosophical or religious questions with the same ease and precision as mathematicians solve number problems—and that it would become a "universal tool."

He made a calculator that could multiply using toothed gears in 1671. Called the Step Reckoner, it did multiplication by repeated addition. Although the Step Reckoner did not use it, Leibniz advocated the use of binary (which is now used in almost all computers) and even imagined a machine that could calculate using binary, using physical objects rather than vacuum tubes or transistors.

Ramon Llul's ideas prefigured technologies that would have been impossible to imagine in the thirteenth century. Thanks to Leibniz, he is venerated today, described as a "prophet of computer science" and the first human being to imagine making logical deductions in a mechanical, not mental, way.

1495

RESEARCHER:
Leonardo da Vinci

SUBJECT AREA:
Automata

CONCLUSION:
Created mechanical automata (and possibly a programmable device)

FANCY DRAWINGS OR FEASIBLE SCIENCE?

DA VINCI'S EXPERIMENTS WITH AUTOMATION

The soaring creative mind of the painter behind the *Mona Lisa,* Leonardo da Vinci, also expressed itself through thousands of pages of notebooks that contained remarkable inventions: from winged flying suits to strange helicopters with screws for rotors. He is thought to have completed relatively few of his inventions, which exist largely as beautiful drawings in the notebooks. But some believe he may have actually constructed one—da Vinci's robot, or mechanical knight.

Renaissance Man
Born in 1452 in the Republic of Florence, da Vinci was a true polymath, in time renowned as a painter, sculptor, architect, and engineer—the ultimate "Renaissance Man." Da Vinci, the illegitimate son of a notary, left school at fourteen when he became a studio boy of Andrea del Verrocchio, a leading Florentine artist. He trained in art, but he wasn't taught Latin and at school learned only limited mathematics. His scientific knowledge in later life largely came from his own observation. A brilliant draftsman and student of human physiology for his art, he applied those skills to the anatomy of machinery.

Many of the mechanical inventions that da Vinci came up with were distinctly warlike, such as a diving suit, with the goal that divers could walk underneath enemy vessels and cut holes in their hulls. Another design showed an armored tank, prefiguring their use in warfare by four centuries. It's perhaps no surprise that his mechanical man took the form of a heavily armored knight.

In 1482, da Vinci moved to Milan to work

as a painter and engineer for Duke Ludovico Sforza, who commissioned his painting *The Last Supper*. Under Sforza's patronage, da Vinci devised his "knight," a machine that, controlled by cables, could wave its arms and open and close its mouth. It had the appearance of a Germanic knight in armor. What is less clear was whether or not da Vinci actually built his mechanical robot knight.

Knight Moves

One theory suggests that the knight was not only built, but also shown off as part of an exhibition by Sforza, possibly as part of a sculpture garden. Mark Rosheim, a roboticist who has produced designs for NASA and Lockheed Martin, and is a collector of da Vinci's sketches, believes that not only did da Vinci build his knight, but that following the design it is still possible to construct the knight today. Rosheim spent five years during the 1990s using da Vinci's detailed drawings of the human body to design for NASA an "anthrobot" that emulated the joints and muscles of the human body. He says that da Vinci's drawings imagine muscles like cables, and that these helped him to match the human body in robotic form.

Rosheim also believes that the knight was fully functional. He said that the robot, "sat up, waved its arms, moved its head via a flexible neck, and opened and closed its anatomically correct jaw— possibly emitting sound while accompanied by automated musical instruments such as drums." Challenged by the BBC in 2002, Rosheim constructed a working model of the "robot" knight that could move its arms, just as predicted.

The Clockwork Lion

Other artists have reconstructed da Vinci automata from the pages of his notebooks. The lion "robot" was recreated in 2009 by Venetian-born automaton designer Renato Boaretto. His clockwork lion, standing more than 4 feet (1 m) tall and 6 feet (2 m) long, could open its mouth, wag its tail, walk and move as if it were roaring. He based his work on a trio of designs for clockwork lions that da Vinci is said to have created for displays. Boaretto also studied other Leonardo manuscripts, including his many studies of clocks, to devise how the lion might have worked—using gears and pulleys, and wound up with a key like a clockwork toy.

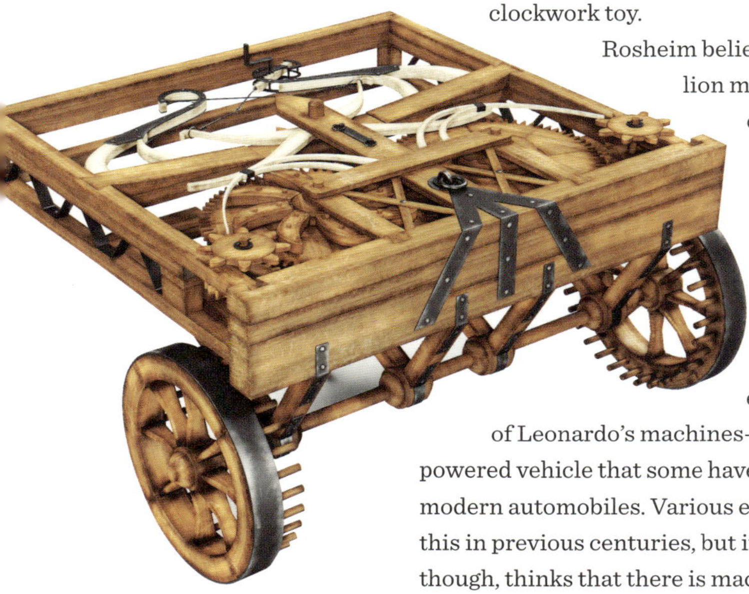

Rosheim believes that gadgets like da Vinci's lion might have been used in displays of automata, similar to Jacques Vaucanson's duck (see page 41) two centuries later.

Self-driving Cart

But da Vinci's experiments in automation might have gone even further. Rosheim examined diagrams for another of Leonardo's machines—his self-propelled cart, a spring-powered vehicle that some have described as an ancestor of modern automobiles. Various experts had attempted to build this in previous centuries, but it had never worked. Rosheim, though, thinks that there is machinery not shown in the diagrams, and that the cart is programmable, which would make da Vinci's design even more visionary.

HOW DO KARAKURI DOLLS WORK?

HOW KARAKURI PUPPETS TAUGHT JAPAN TO LOVE ROBOTS

1600s
RESEARCHER:
Takeda Omi
SUBJECT AREA:
Clockwork puppets
CONCLUSION:
Japanese people welcomed clockwork "robots" into their homes

The blank-faced doll rolls forward with a cup of tea on a tray. When a guest picks up the cup of tea, it stops and waits patiently, and when the empty cup is replaced, the doll trundles away politely, nodding its head.

The tea serving robot is a uniquely Japanese invention: it's one of several forms of karakuri puppet, machines used for centuries on stage and in wealthier Japanese homes as a novelty.

Known as *karakuri ningyo* (karakuri puppets), the dolls date back to Japan's Edo period (1603–1868), when Japanese craftsmen first adapted Western clockmaking technology to create strange, lifelike dolls, at first for the theater stage.

The popularity of Karakuri puppets over the centuries may help to explain Japan's long love affair with robots, with robot pioneers such as Yoshiyuki Sankai speaking of how people in Japan see robots more optimistically than people in the West (see page 79), and many innovations in consumer robotics coming from Japanese companies.

In *Japan at Play*, a history of the culture of Japan, anthropologist Joy Hendry writes, "The automaton puppet was a prototype of the robot [...] flourishing in industrial Japan today. We can say that the Japanese learned to tame the machine by means of the karakuri puppet."

Like Clockwork

The history of karakuri dolls dates from Japan's first clock, dedicated by sixteenth-century Jesuit missionary

Francis Xavier to Ouchi Yoshitaka, Lord of Suho. Local craftsmen were quick to adapt and reverse-engineer the clockwork technology, using it for their own purposes.

Takeda Omi, a local businessman and impresario in Osaka, used clockwork, along with other trickery such as stoppers and gears, to animate dolls that performed on a stage in Osaka's Dotonbori entertainment district. The dolls included devices powered by the water in the canal, and featured stylized characters who would do handstands on trapezes.

Writer Ihara Saikaku enthused that Takeda had "fabricated a wheeled mechanical doll with a main spring that can move in any direction. It holds a teacup. The movements of the eyes, mouth, and feet, the motion of the extending of the arms, as well as its bowing gesture are remarkably lifelike."

The dolls became an attraction in Osaka, and the karakuri business lived on through the Takeda family. People said, "You haven't seen Osaka if you haven't been to the Takeda Karakuri."

East Versus West

The Takeda troupe made a visit to Edo in 1741, returning again in 1757. The first show was titled "Ten Months in the Mother's Womb" and featured a puppet of a three-month-old baby, which played the flute and defecated on stage—not dissimilar to Jacques Vaucanson's defecating duck, which drew paying audiences in eighteenth-century France (see page 41). Other dolls in the Takeda stage shows represented deities, demons, and skeletons. The movements of karakuri dolls are thought to have influenced the stylized way human Japanese actors move in traditional theater.

But the dolls were not long confined to the stage. Other, larger Karakuri would stand on street floats in religious festivals. Zashiki karakuri, or "room" dolls, were designed for rich people to entertain guests, and were adopted by feudal lords and other dignitaries as party tricks.

Most popular among these were the "chahakobi ningyo," which

used clockwork and hidden wheels to serve tea to a guest and return to the host. Popular tricks included being able to specify the distance to which the doll would roll out, before it would "serve" tea to the guests. The cogs inside Karakuri dolls are often made of wood, handmade by craftsmen, and traditionally the spring inside the doll is made from a whale's baleen—the bristly comb through which some whales filter food in their mouths. Purists claim that modern recreations with springs made from metal or plastic cannot recreate the subtle movements of traditional karakuri.

Springing to Life

There's a direct connection between the creators of karakuri puppets, and Japan's hi-tech industries today (Japan became one of the most enthusiastic adopters of Unimate industrial robots, see page 96).

Tanaka Hisashige, the founder of the business that later became the Japanese company Toshiba, was a famous creator of karakuri dolls (including a doll that shot arrows and one that wrote a letter) in his teens, before going on to invent technological innovations including lights, which earned him the title of "Japan's Thomas Edison."

Born in 1799, he built mechanical versions of Karakuri, using hydraulic pressure, gravity, and air pressure: one, Yumihiki Doji, "arrow-shooting boy," uses a clockwork mechanism, thirteen threads with levers and twelve moving parts, which picked up four arrows and shot them at a target—being "programmed" so one would always miss. Tanaka toured the country with his dolls and became a celebrity in his own right, before moving to Tokyo to develop a telegraph system for the government.

Today, karakuri dolls are shown off in exhibitions and as displays for tourists—and Japan's culture is uniquely welcoming to robots, with robot receptionists, robots working in nursing homes, and pioneering robot technology companies such as Cyberdyne (see page 145). Japan's government has invested heavily in robotics. *The Japan Times* recently suggested that Japan's aging population wants "automation, not immigration" in the workforce.

CHAPTER 2: Industry and Automation 1701–1899

The dawn of the Industrial Revolution saw innovation blossom in both ideas and the first "automated" machinery. From the first agricultural equipment with moving parts to looms that wove pictures so detailed they fooled observers into thinking they were paintings, machines became "programmable"—and new inventions such as the Jacquard loom, controlled by computer-style punch cards, became so valuable that their export from France was forbidden by Napoleon.

Other visionaries such as Thomas Bayes would lay the foundations of data science with ideas on probability, which would become important in robotics more than a century later. Meanwhile, inventor Charles Babbage would dream of two computing machines that would never be built in his lifetime, and his collaborator Ada Lovelace would write the first computer program, also for Babbage's nonexistent machine.

1701

RESEARCHER:
Jethro Tull

SUBJECT AREA:
Agricultural automation

CONCLUSION:
Created the first agricultural machine with moving parts

HOW CAN SOWING BE MORE EFFICIENT?

HOW JETHRO TULL'S SEED DRILL BROKE NEW GROUND

To modern eyes, an eighteenth-century horse-drawn seed drill doesn't look like a device that might light one of the first sparks for the age of automation. But the seed-planting contraption first tested on a British farm near Hungerford, England, would change agriculture forever. It also paved the way for machines that would take instructions.

Local farmer Jethro Tull went on to be described as "the greatest improver" that agriculture has known, and his inventions carved out a path for further innovations in the Industrial Revolution.

His seed drill was the first agricultural machine with moving parts, and improved efficiency as well as saving labor. But Tull also championed some very strange ideas about farming, and faced considerable opposition to his inventions (with some later historians describing him as a "crank").

Pipe Dream

Born in 1674, Tull had studied the pipe organ and trained as a lawyer before returning to his family farm. Frustrated by his farm's inefficiency, he invented his seed drill as a labor-saving device. Until the invention of his machine, seed had been scattered by hand into furrows (known as "drilling"), leading to a huge amount of wastage.

Tull instructed his workers to "drill" at precise, low densities, but they were unwilling to learn new ways of working. John Donaldson wrote in his 1854 work *Agricultural Biography* that Tull, "experienced the usual difficulties that attend all new undertakings . . . the old implements were unsuitable and clumsy; the workmen were awkward and unwilling."

By 1701, Tull was so frustrated with his workers that he invented

a machine to do the work instead. Inspired by a pipe organ that he had seen dismantled, his drill had a rotating cylinder to guide seeds from the hopper to a funnel, where they would drop straight into a channel dug by a plow at the front. As the machine passed, a harrow would cover over the seed with soil automatically.

Tull's invention started as a one-man device for sowing seeds in one row at a time, but he upgraded it to sow in three uniform rows, drawn by horses. "In drilling, seed lies all at the same, just depth, none deeper, none shallower than the rest," he wrote. "There's no danger of the accidents of burying or being uncovered, and therefore no allowance must be made for them."

Prosperous Times

The invention saved up to a third of the seed being used, making his Prosperous Farm more profitable, although Donaldson noted that there was reluctance among the workforce to adopt this new technology, saying that the workers "would break the new implements in order to continue the lazy working of the old ones."

A mechanical horse-drawn hoe, another Tull invention, helped to remove weeds from between the lines of crop plants, increasing

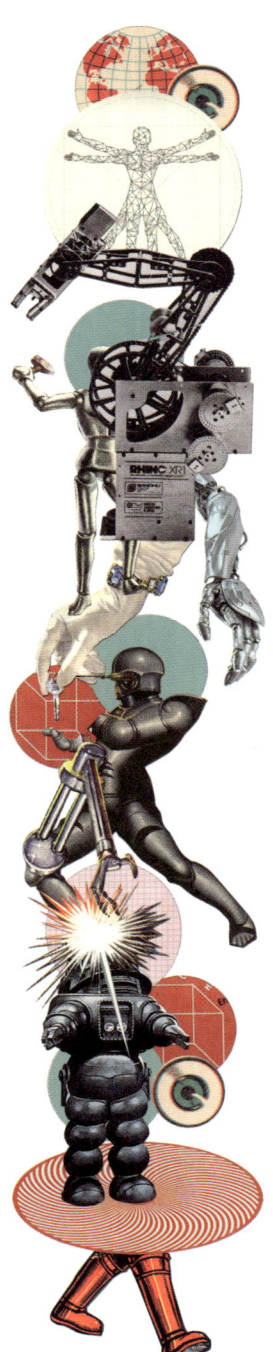

the efficiency of farmland even further. Traveling in France and Italy, he was also impressed with cultivation methods used in vineyards, where the earth between the rows of vines was broken up, improving the plants' access to water, and reducing the need for manure.

Strange Ideas

Inspired, Tull published a book in 1731, titled *Horse-hoeing Husbandry Or, An Essay on the Principles of Vegetation and Tillage. Designed to Introduce a New Method of Culture*. There was considerable resistance to his ideas, not helped by the fact that his sensible ideas such as his seed drill sat alongside completely unsound notions, such as the fact that soil itself was nourishing for plants, without any need for manure. Tull believed that the soil would "feed" the plants, as long as it was sufficiently broken up. He was completely wrong.

He wrote: "All sorts of dung and compost contain some matter which, when mixed with the soil, ferments therein; and by such ferment dissolves, crumbles, and divides the earth very much. This is the chief and almost only use of dung . . ." His wrong-headed insistence that manure was not necessary and that breaking up the soil was sufficient, permeates his book.

Tull died in 1741, and his ideas, both good and bad, were not widely accepted during his lifetime. His seed drill was too expensive for most farmers in the 1800s, but was modified and improved by others over the next century. During the nineteenth century, his invention was popularized by James Smyth and his sons, who used new casting technology to produce a cheaper, more efficient version of Tull's drill, which was then exported all over Europe.

Tull's "scientific" approach to agriculture was also widely influential. Donaldson wrote, "The name of Tull will ever descend to posterity as one of the greatest luminaries, if not the very greatest benefactor, that British agriculture has the pride to acknowledge. His example furnishes the vast advantages of educated men directing their attention to the cultivation of the soil."

WHAT WILL HAPPEN NEXT?

HOW BAYES'S THEOREM LETS US PREDICT THE FUTURE

1763

RESEARCHER:
Thomas Bayes

SUBJECT AREA:
Probability

CONCLUSION:
Bayes's Theorem lets us predict outcomes from what happened before

How do we work out what is likely to happen next? Oddly enough, the way we think about probability today was shaped by a clergyman's argument about the existence of God and whether it is reasonable to believe in miracles like Jesus's Resurrection.

Thomas Bayes's idea, known as Bayes's Theorem, helps people forecast outcomes by basing them on previous data, and is used in everything from machine learning to tests for Covid-19. It's a powerful idea because it can take into account, for example, a test that is sometimes inaccurate, or an unreliable witness, and deliver a probability based on all the variables.

The theorem is a simple method to calculate, working on how often something has occurred in previous trials, and so how often it will occur in future ones. It's used widely in finance and in developing new drugs, and is increasingly important in artificial intelligence.

Bayes was a mathematician, Presbyterian minister, and theologian who was born in London in 1702. During his life, he worked on calculus, and was a member of the Royal Society.

But his most famous work, *Essay Toward Solving a Problem in the Doctrine of Chances*, was published posthumously by his friend, Welsh philosopher and mathematician Richard Price, in 1763. Price's motive in publishing the work was partly due to a desire to prove the existence of God.

Philosopher David Hume had written in his 1748 essay *Of Miracles* that merely observing a miracle was not enough to prove that it had

$$P(A|B) = \frac{P(B|A)P(A)}{P(B)}$$

happened. Hume wrote, "No testimony is sufficient to establish a miracle, unless the testimony be of such a kind, that its falsehood would be more miraculous than the fact which it endeavors to establish."

Hume's essay was generally taken as an attack on religious belief, and Price was determined to refute him using Bayes's mathematics.

Calculating God

In Price's introduction to Bayes's essay, he chose the example of a tide flowing in, which has been observed to come in at the same time of day a million times. Price used Bayes's Theorem to calculate that the odds that it will not appear one day are not (as some might imagine) a million to one, and that there is a roughly fifty percent chance they could be as little as one in 600,000.

Price wrote, "Suppose that universally a person was to reject all accounts which he reads or hears of facts, what would be thought of such a person? How soon would he be made to see and acknowledge his own folly?"

In the context of the Resurrection, Price's point (using Bayes's idea) was that there are multiple accounts by independent witnesses, which alters the probability.

Statistician and historian Stephen Stigler wrote, "Hume underestimated the impact of there being a number of independent witnesses to a miracle, and that Bayes's results showed how the multiplication of even fallible evidence could overwhelm the great improbability of an event and establish it as fact."

Calculating Probability

Bayes's Theorem is expressed as:

$$P(A|B) = \frac{P(B|A)P(A)}{P(B)}$$

$P(A|B)$ is the probability of A occurring given that B is true
$P(B|A)$ is the probability of B occurring given that A is true
$P(A)$ and $P(B)$ are the probabilities of A and B occurring

So for instance, if you draw a card from a deck of fifty-two cards, the probability that the card is a king is four divided by fifty-two, which comes out as 7.69%, or one in thirteen. But if someone looks at the card and sees it is a face card, we can calculate this using the formula since we know that the probability of a king being a face card is 1/1. The probability that it is a king (knowing that the card is a face card) is 33% because there are twelve face cards in a deck.

Bayes and Covid-19

Bayes's Theorem was widely used in dealing with the coronavirus pandemic—and explains some of the more counterintuitive results of lateral flow tests (the relatively inaccurate rapid tests used in workplaces and schools). With lateral flow tests, the probability of getting an incorrect, positive result when you are not infected is roughly one in 1,000.

But when there's a low rate of infection in the population, a relatively high number out of the positive results given will be false positives (which is one of the reasons people often had to take a more accurate PCR test to confirm the results of a lateral flow test). It's a counterintuitive result, and something that Bayes's Theorem helps to explain. Bayesian thinking is also crucial to vaccine trials.

Today, Bayes's Theorem is essential in machine learning and artificial intelligence, allowing scientists to assess the probabilities that something is true based on new evidence. It's been described as "the most important formula in data science," and has helped scientists to do everything from improving mobile phone signals, to filtering email spam, to predicting the weather. In robotics, Bayes's Theorem is used to calculate the probabilities of a robot's next step, given the steps the machine has already executed.

During his lifetime, Bayes never achieved fame for his theorem, but in the twenty-first century, his ideas have never been more popular. In 2020, following revelations about Sir John Cass's links to the slave trade, the London business school that bore his name was renamed to the Bayes Business School.

1804

RESEARCHER:
Joseph Marie Charles (known as Jacquard)
SUBJECT AREA:
Automation
CONCLUSION:
Jacquard's punch cards changed textiles forever, and inspired early computers

CAN A MACHINE TAKE ORDERS?

THE DAWN OF PROGRAMMABLE MACHINES

At one of computer pioneer Charles Babbage's parties, he showed off a picture on his wall to the Duke of Wellington and Prince Albert, Queen Victoria's husband. The Duke asked whether the detailed portrait of Joseph Marie Charles (known as Jacquard to distinguish him from other branches of the same family) was an engraving. Prince Albert (who had seen a previous example of the image) replied, "It is not an engraving."

The portrait was woven, and had been designed to show off the capabilities of Jacquard's invention, an automated loom. As Babbage wrote, it was a "sheet of woven silk, framed and glazed, but looking so perfectly like an engraving, that it had been mistaken for such by two members of the Royal Academy."

The portrait had 24,000 rows of weaving, all precisely controlled by the punch cards of Jacquard's loom. It was based on a painting of Jacquard by Lyons artist Claude Bonnefond, and designed to show off the unheard-of precision of the machine, a programmable loom.

Babbage would go on to use similar punch cards in his designs for his Analytic Engine, which prefigured today's digital computers by more than a century (see page 43).

Programming Pictures

Joseph Marie Charles, born in 1752, was a weaver's son who had been through bankruptcy and the French Revolution, in which he had fought in the defense of his hometown, Lyon.

Before the invention of the Jacquard loom, even the most experienced two-man team of weavers could only produce one inch of detailed fabric per day. Weavers had to manually adjust each of 2,000 threads on the drawloom (a design that had remained largely unchanged since the second century), with a weaver working alongside a "drawboy" who would sit inside the loom and adjust the threads for each row according to the directions of the weaver. Even the most experienced teams could not manage more than two rows per minute. Jacquard's invention was to reshape the textile industry worldwide.

The Defecating Duck

He was not the first person to attempt to make an automated loom. Jacques de Vaucanson, who was appointed inspector of French silk factories in 1741, invented an automated loom, where instructions were "stored" on a metal cylinder, like the ones in children's music boxes.

Vaucanson also merits a mention for his devotion to automata, which became fashionable in eighteenth-century France. Voltaire famously described him as "A rival to Prometheus," who "seemed to steal the heavenly fires in his search to give life."

Most picturesquely, Vaucanson had made his own robotic duck, which could quack, flap its wings, and eat—as well as passing out waste. It had a preloaded tank of waste, which would slop out as it ate. The inventor showed off his defecating duck in a hall in Paris in the winter of 1738, alongside two humanoid flute players and pipe players. But Vaucanson's loom was considerably less successful than his duck. The cylinders were too expensive to produce, and it was discontinued.

Pleased as Punch

Jacquard's loom, though, was different: it was controlled by a set of punch cards and hooks: each row of holes punched into the cards corresponded to a row of thread. The hooks passed through holes with a thread, creating a pattern. Complex patterns required a deck of cards.

Jacquard arrived at his invention slowly, patenting one loom in 1800, under the title, "machine designed to replace the drawboy in the manufacture of figured fabrics." By 1804, the loom came to the attention of Napoleon, who granted Jacquard a pension for life, and a fee for every loom sold.

Legend also has it that, around this time, Jacquard was thrown into a river by angry weavers fearful of losing their livelihoods—although this seems unlikely if we are to believe the glowing description that his biographer The Count of Fortis wrote of him: "Jacquard was a man who was most at home among workmen. He was happiest in their company and to know him as he really was one had to see him in his ordinary clothes in a weaver's studio, giving the weavers instruction on how to make the best use of the loom."

His loom became central to Napoleon's ambition to compete with British industry and exports to Britain were banned. Naturally, some were smuggled out of France (one in a barrel of fruit), forming the foundation of the silk industry around the world.

Jacquard's punch cards were to be even more influential. In the late nineteenth century, American Herman Hollerith began to use punched cards in his "tabulator" machines to record census data. Hollerith's company would eventually (via a series of mergers) become computer giant IBM and the cards became the major medium for storing and sorting data for the first IBM computers.

By the late 1960s, America used 500 billion punch cards per year, up to 400,000 tons of paper. Even in the late 1990s, some companies still used punch cards, the descendants of Jacquard's invention, for operations such as payroll data.

HOW DID MATHEMATICS FIND ITS ENGINE?

BABBAGE, LOVELACE, AND THE COMPUTING MACHINE

1832
RESEARCHER:
Charles Babbage and Ada Lovelace
SUBJECT AREA:
Computing
CONCLUSION:
Designed (but never built) a device like a modern computer

Mathematicians Charles Babbage and Ada Lovelace are best known for two devices that are widely considered the great-grandparents of every computer on our planet today—the Difference Engine and the Analytical Engine, even though neither was fully built in their lifetime.

Babbage completed only small parts of the first machine, the Difference Engine, with a "beautiful fragment" of his calculating device delivered in 1832, shortly before financial disaster struck.

But today it's clear that he and his collaborator Ada Lovelace—daughter of poet Lord Byron and mathematician Lady Byron—were visionaries who, through an algorithm, worked out a way of using the machine to calculate Bernoulli numbers (a mathematical sequence). This is widely considered as the first computer program.

Calculated by Steam
At first, Babbage focused on creating a machine that could do mathematical calculations. Born in 1791, the banker's son taught math at Cambridge and became influential in advancing the sciences in Britain. In 1821, while examining handwritten mathematical tables with his friend, the astronomer John Herschel, Babbage was frustrated to find error after error. "We commenced the tedious process of verification," he later wrote. "After a time, many discrepancies occurred, and at one point these discordances were so numerous that I exclaimed, 'I wish to God these calculations had been executed by steam!'"

Rather than steam power, Babbage settled on clockwork for the designs for his Difference Engine, using brass gear wheels, rods, pinions, and ratchets. Within the machine, numbers were represented by the positions of ten-toothed metal wheels.

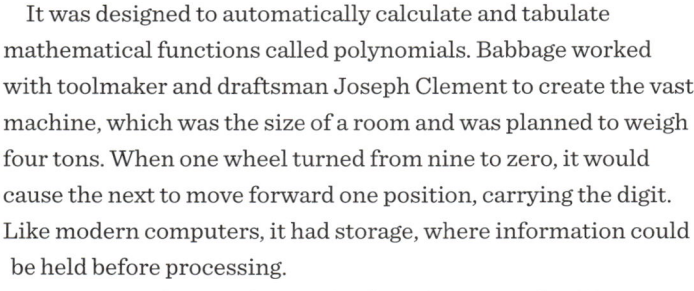

It was designed to automatically calculate and tabulate mathematical functions called polynomials. Babbage worked with toolmaker and draftsman Joseph Clement to create the vast machine, which was the size of a room and was planned to weigh four tons. When one wheel turned from nine to zero, it would cause the next to move forward one position, carrying the digit. Like modern computers, it had storage, where information could be held before processing.

But work ground to a halt after Clement and Babbage argued over money for moving Clement's workshop to Babbage's house. Part-funded by government grants, the Treasury had already spent £17,500 on the Difference Engine. Funding was withdrawn.

Punch Card Power

In his later years, Babbage planned an even more ambitious machine, the Analytical Engine, which would have been controlled by punch cards, and similar in many ways to today's computers. It had a memory (the "Store") and a central processor (the "Mill") and ways to input and output data.

"As soon as an Analytical Engine exists, it will necessarily guide the future course of the science," he said. "Whenever any result is sought by its aid, the question will then arise—by what course of calculation can these results be arrived at by the machine in the shortest time?"

Fellow mathematician Ada Lovelace wrote that the Analytical engine "weaves algebraic patterns, just as the Jacquard loom weaves flowers and leaves." She also wrote a translation of Luigi Menabrea's French account of Babbage's work, in which her own appendices and notes made up almost two-thirds of the final text. She published her work in Taylor's *Scientific Memoirs* in 1843 under the initials A.A.L., culminating in an explanation of how to use Babbage's engine to calculate the Bernoulli numbers. Lovelace offered a detailed explanation of how the various parts of the device would do this, stating that it was an "illustration of the powers of the engine."

She also envisioned the idea of a loop in a computer program (where a program repeats an action until a certain condition is achieved), which she likened to a "snake biting its tail," and imagined a future where Babbage's machines might compose music—something that has become reality today, with composers using software to generate music, including David Cope's pieces in the style of Mozart.

"Silent Witness to Great Hopes Dashed"

Babbage invented many things during his life, including cowcatchers for trains—the plow-shaped device attached to the front of the locomotive to deflect obstacles—and was the first to realize that you could "read" former weather patterns using rings in tree trunks. He also passionately believed that new inventions should be freely available to the public. When he built the first ophthalmoscope in 1847, he refused to patent it. Naturally, several years later, someone else did.

Modern experts have tended to point to a certain lack of focus as to why Babbage never built any of his computing machines, saying that when a new idea overtook him he would bound away after that, forgetting what he was supposed to be working on. Others have suggested that the limitations of materials available to Babbage made his machines difficult to produce. It's certainly the case that the "beautiful fragment" of the Difference Engine represented a leap forward in precision engineering.

Babbage's obituary in *The Times* had a distinctly mocking tone: "No sooner did Mr. Babbage, like an honest man, communicate the fact [of needing more money] to the Government than the then Ministers, with Sir Robert Peel and Mr. H. Goulburn at the head of the Treasury, took alarm, and, scared at the prospect of untold expenses before them, resolved to abandon the enterprise." The obituary went on to describe how hundreds of pages of plans for Babbage's devices had gone on show, alongside the small parts of the machine Babbage had successfully built.

More than 100 years later, experts at London's Science Museum built a real-life version of the Difference Engine, using the same technology Babbage envisioned. The five-ton mechanical machine worked perfectly, just as Babbage had described.

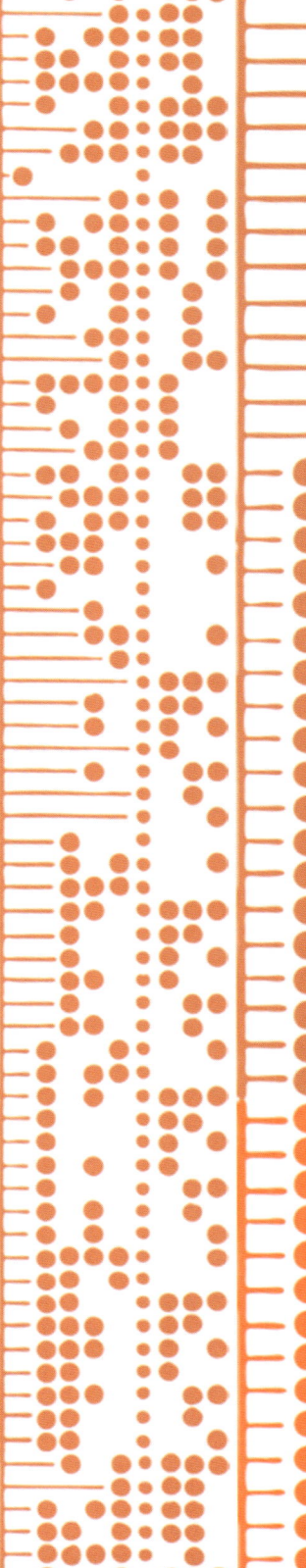

1871

RESEARCHER:
Richard March Hoe
SUBJECT AREA:
Automation
CONCLUSION:
Automation paved the way for the modern newspaper era

HOW DID MECHANIZATION CHANGE PUBLISHING?

HOE'S LIGHTNING PRESS

At the dawn of the nineteenth century, printing was still fundamentally the same as it had been when the Gutenberg press was invented in 1440, with letters arranged in a tray, covered in ink, and paper pushed down on top.

Johannes Gutenberg's invention had sparked a publishing revolution, with the device capable of producing 3,600 pages per day, meaning that 200 million books were in print by the sixteenth century.

The Gutenberg press had helped to catalyze the original "information age," the Renaissance, but a new technology, built around speed and automation, would vastly increase newspaper readership in the U.S. and Britain, setting the stage for the modern world. American printing pioneer Richard March Hoe would be central to the change, creating a machine (his "web perfecting press"), which accelerated newspaper production to the speeds necessary to inform millions of people at once.

Rotating Presses

The change began with a shift to rotating presses, led by newspapers in Britain such as *The Times*, and based on machines designed by Friedrich Koenig and Andreas Bauer. When John Walter, the owner of *The Times*, first used a steam-driven Friedrich Koenig cylinder printing machine in 1814, he kept it secret from his staff, fearing a repeat of the Luddite machine-breaking seen in the textile industry, where activists had attacked factories and broken machines in fear of losing their jobs.

In 1812, Parliament had made machine-breaking a capital offense. Factory owners of the period also built hidden chambers so they could remain safe if their premises were attacked by disgruntled workers.

To avoid such attacks, Walters told staff that he was holding the presses for a big story, and printed the entire edition in secret. Staff who were displaced by the new technology were kept on full wages until they found new jobs.

The next day's edition proudly announced, "The reader of this paragraph now holds in his hand, one of the many thousand impressions of *The Times* newspaper, which were taken off last night by a mechanical apparatus. A system of machinery almost organic has been devised and arranged, which, while it relieves the human frame of its most laborious efforts in printing, far exceeds all human powers in rapidity and dispatch."

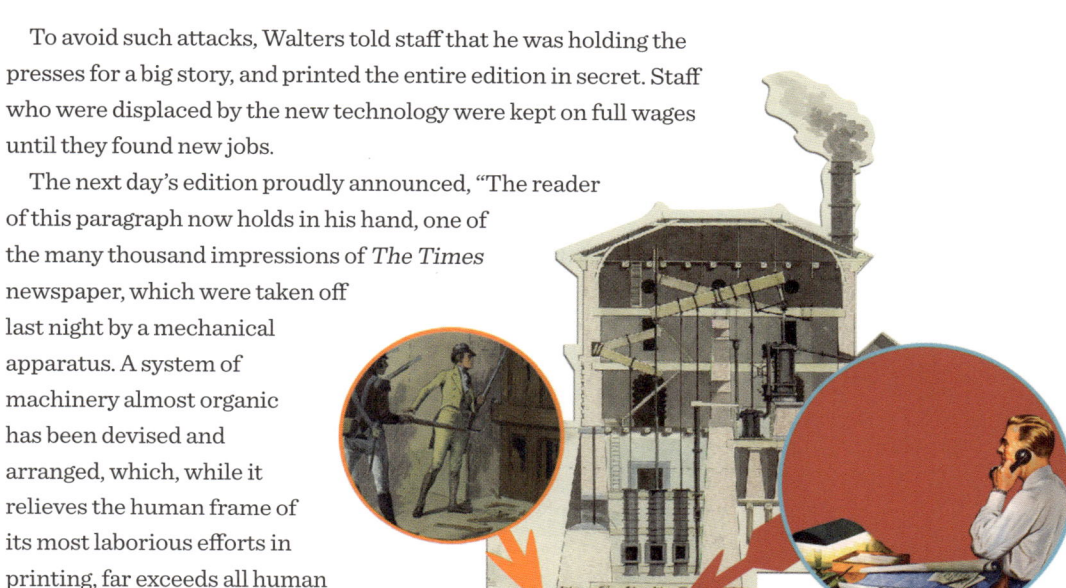

Koenig himself wrote in *The Times* days later of how his new machine could print 800 pages per hour, and how his machine had succeeded, where previous attempts had seen "thousands of pounds" invested with no result.

Lightning Strikes

The United States was slower to adopt such machinery, but would soon take the lead in the rapidly evolving and ever-more-automated printing business. Born in 1812, Richard March Hoe inherited a printing business from his father, building on his innovations to create his own refinement of the rotary printing press, which would revolutionize the newspaper industry.

Granted a patent in 1847, Hoe's Lightning Press raised speeds, with the type on a revolving cylinder, surrounded by four iron cylinders each carrying sheets of paper. The design allowed thousands of papers to be printed per hour, with more cylinders being added to create ever faster machines.

In *Great Fortunes and How They Were Made*, the nineteenth-century writer James McCabe wrote, "The ten-cylinder press costs fifty thousand dollars, and is regarded as cheap at that immense sum. It is one of the most interesting inventions ever made. Those who have seen it working in the subterranean press-rooms of the journals of the great metropolis will not soon forget the wonderful sight."

Age of Newspapers

Powered by steam, the machines not only offered existing newspapers the chance to vastly increase their circulation, but also meant new newspapers launched. Robert Hoe, Richard Hoe's nephew wrote, "A revolution in newspaper printing took place. Journals which before had been limited in their circulation by their inability to furnish the papers rapidly increased their issues, and many new ones were started. The new presses were adopted not only throughout the United States, but also in Great Britain. The first one put up abroad was erected in 1848, in the office of 'La Patrie' in Paris."

But Hoe (who rapidly became a wealthy man as his invention was adopted worldwide) was to refine the printing press still further, with machines even closer to today's high-speed printing. His "web" printing machine advanced the idea. It printed from a single roll of paper five miles long and could print thousands of copies of papers on double sides in just seconds.

His nephew described the operation of the machine, "As the paper unwinds it passes first over a jet of steam which slightly dampens and softens its hard surface and fits it for receiving impressions, without leaving it wet or sodden. It passes under a plate cylinder, on which are thirty-two curved plates, inked by seven large rollers, which print thirty-two pages on one side. Then it passes around a reversing cylinder which presents the other side of the paper to another plate cylinder.... This is done quickly (in less than two seconds) but with exactness."

The roll passed over knives that cut the paper apart, delivering a completely printed, folded newspaper. The machine could produce 18,000 finished newspapers per hour, and paved the way for the era of mass newspaper circulation around the world.

WHO INVENTED REMOTE-CONTROLLED VEHICLES?

NIKOLA TESLA'S "TELEAUTOMATON"

1898

RESEARCHER:
Nikola Tesla
SUBJECT AREA:
Radio-controlled drones
CONCLUSION:
Showed off the potential of driverless boats, planes, and cars

Drones often seem like a uniquely twenty-first-century phenomenon, from the whirring toys beloved of hobbyists to the deadly weapons that fly thousands of feet above war zones (see page 130).

But strangely enough, the first "drone" was actually shown at the tail end of the nineteenth century in New York (although no one bar its inventor realized its commercial potential at the time).

Serbian engineer Nikola Tesla showed off what he described as a "teleautomaton," a three-foot-long, battery-powered model ship in a tank of water, controlled via radio waves.

Born in 1856, Tesla had moved to America and was at that point making a good living as an inventor of, among other things, alternating current, the system used in utility power today. His innovations in electricity are one of the reasons the car company Tesla bears his name.

In 1898, he had been granted a patent for a "Method of and Apparatus for Controlling Mechanism of Moving Vessels or Vehicles." He demonstrated his control over the machine by asking it questions and having lights on board the machine flash the correct number of times. He later said, "When first shown [...] it created a sensation such as no other invention of mine has ever produced." He sent signals to the boat via a small box with control levers on the side.

Monkey in the Machine

At the demonstration, the technology was considered so outlandish that several people were convinced that Tesla was cheating in some way, perhaps by controlling the vessel with his mind. Others thought that there was a tiny monkey inside the vessel, steering it in response to his commands.

In his patent, Tesla proclaimed that he had invented, "certain new and useful improvements in methods of and apparatus for controlling from a distance the operation of the propelling engines, the steering apparatus, and other mechanism carried by moving bodies or floating vessels," and said that it could be applied to "boats, balloons, or carriages."

War Machines

Tesla's demonstration took place five years before the Wright brothers achieved powered flight, and he predicted that his invention would see wide use in exploration and even in hunting whales. Perhaps most prophetically of all given the widespread use of drones in warfare today, he was also confident that wireless remote-controlled machines would form a weapon so deadly that it would end all wars.

Tesla wrote in his patent application that weapons based on his "teleautomaton" would be so lethal that it would persuade nations to abandon war altogether. "The greatest value, for by reason of its certain and unlimited destructiveness it will tend to bring about and maintain permanent peace among nations," he wrote.

Tesla was highly prescient in this: in the decades since, military applications have been central to driving drone technology (see page 130). Experimental wireless remote-controlled aircraft flew in World War I, and by the time of the Vietnam War, surveillance drones were an important part of warfare. Many of the experts behind self-driving cars cut their teeth in the DARPA Grand Challenge, a competition designed by the U.S. military to come up with vehicles to help soldiers on the battlefield (see page 142).

Prophet, not Profit

Tesla never made money from his invention: despite the public's interest, he failed to persuade investors that there was any value in his "teleautomaton" and the U.S. Navy did not take an interest either.

In the years after his invention, other inventors showed off remote-control devices: Spanish engineer Leonardo Torres-Quevedo operated a boat from a mile's distance near Bilbao in front of an astonished crowd in 1905, controlled by his Telekino, which is often described as the first remote control. Torres-Quevedo would go on to invent a chess-playing automaton that in many ways started the journey toward artificial intelligence.

It would not be the last time Tesla would fail to profit from his ideas. The engineer was something of an eccentric, having also claimed that he was working on a mysterious death ray that could shoot down thousands of planes at once, as well as a car that ran on cosmic rays, and a machine to photograph people's thoughts.

But Tesla was a true visionary. In an interview in 1926, he predicted the smartphone revolution with startling accuracy. "When wireless is perfectly applied the whole earth will be converted into a huge brain. We shall be able to communicate with one another instantly, irrespective of distance," he said. "Not only this, but through television and telephony we shall see and hear one another as perfectly as though we were face to face, despite intervening distances of thousands of miles; and the instruments through which we shall be able to do this will be amazingly simple compared with our present telephone. A man will be able to carry one in his vest pocket."

CHAPTER 3: The Dawn of Modern Robotics 1900–1939

The first half of the twentieth century saw the invention of the words "robot" and "robotics": one by Czech playwright Karel Čapek, the other by the prolific, bushy-bearded science fiction writer Isaac Asimov. The two had very different visions of robots, with Čapek's play a nightmare vision of a future where robots wipe out humanity—and Asimov imagining a peaceful future where robots live with human beings, governed by his "Three Laws of Robotics."

As the idea of robots took root in novels, plays, and films such as Fritz Lang's iconic *Metropolis*, technology evolved too, with the first robots built to do human jobs, and a pioneering robot arm that was too advanced to find work. Meanwhile, in the ruins of wartime Berlin, a computing pioneer worked on a machine that would be destroyed by Allied bombs, and that would remain unknown to the outside world until the fall of the Third Reich.

1914

RESEARCHER:
Leonardo Torres-Quevedo

SUBJECT AREA:
Chess AI

CONCLUSION:
The first computer game (but one where you could only lose)

COMPUTER VS. MAN?

THE FIRST (UNBEATABLE) CHESS-PLAYING AUTOMATON

If most of us were asked to imagine the dawn of computer games, we'd probably picture young people playing arcade games such as Space Invaders in the 1970s, or perhaps scientists huddled around huge mainframe computers in the decades before that. But a machine commonly described as the first computer game first took on human beings in 1914. Not only that, it never lost.

Unlike previous "chess playing automatons," El Ajedrecista, designed by civil engineer Leonardo Torres-Quevedo, was not a cheat. It was the latest in a line of machines designed by the prolific Spaniard, including calculating machines that had solved algebraic equations.

Born in 1852 and independently wealthy, he had traveled widely in Europe before settling down to become a full-time inventor. Among his patents and inventions were funicular railroads, airships, and cable cars—and what is often described as the first remote control (for controlling dirigible balloons from the ground), although he wrote that he imagined it could be used for many different mechanisms. His inventions included the Whirlpool Aero Car over Niagara Falls, which was completed in 1916 and still works today.

In an essay, "Automatics. Its Definition. Theoretical Extent of Its Applications," he explained that he had built the chess player to demonstrate his idea that machines could substitute for humans in areas that had previously been solely reserved for human intelligence.

Mechanical Men

In previous centuries, various "machines" had been presented that were supposedly capable of playing chess, most famous among which was the Mechanical Turk, shown off by Wolfgang von Kempelen to impress Empress Maria Theresa of Austria in 1770.

The wooden man would spring to life, grasping chess pieces

and playing surprisingly strongly, beating several human players. Audiences at the time thought it may have been controlled by evil spirits or even a chess-playing monkey.

In fact, there was a human being hidden inside it, playing the pieces from beneath the board, using a mechanism called a pantograph that was later to become important in twentieth-century robot arms.

There was no such trickery to Torres-Quevedo's machine. The machine was electromechanical, making simple "decisions" on what move to play next. It only played a simple chess endgame, with a king and rook playing against the player's king. It didn't always make the best move, and occasionally games would last for more than fifty moves, but in the end it would always checkmate its opponent.

"This Apparatus Has No Purpose"

It marked a first step into artificial intelligence, and was the first machine built to follow rules with a defined end goal (known as "heuristics"), a technique now used to help artificial intelligence algorithms find answers. The machine followed a set of conditional rules, which meant that it could always win. Torres-Quevedo said in an interview, "This apparatus has no practical purpose, but it supports the basis of my thesis: that it is always possible to create an automaton the actions of which depend on certain conditions and which obey certain rules that can be programmed when the automaton is being produced."

Scientific American reported on the machine with breathless excitement, saying that Torres-Quevedo "would substitute machinery for the human mind." The magazine described how El Ajedrecista could detect moves that were against the rules, lighting lamps on its base to protest. If three lamps lit up, the game

was over. "The novelty in the matter is that the machine looks over the field and selects one possible action in preference to another. There is, of course, no claim that it will think or accomplish things where thought is necessary, but its inventor claims that [...] the automaton can do certain things which are popularly classed with thought."

Theory of Automatics

The first incarnation of the machine was improved on in 1920, when Torres-Quevedo built a second version of El Ajedrecista, where, rather than a mechanical arm moving electrical jacks on an electrical-looking board, the pieces "move themselves" across a normal-looking chess board, guided by electromagnets beneath.

It was also equipped with a voice (in the form of a gramophone attached to the machine). When the machine had its opponent in check, it would announce, "Jaque al rey," (check) and when the game was over, it would announce, "Mate," (checkmate).

In 1920, Torres-Quevedo also showed off an arithmometer that could solve sums typed in via a typewriter to an audience in Paris. The machine calculated the results of mathematical formulas, using an electromechanical device with electromagnets, switches, and pulleys.

The machine would output its answer via another typewriter, an innovation that prefigured the way computers were used throughout the twentieth century. The two typewriters were connected by an electrical cable, so in theory, they could have been located in different places.

To use the machine, the operator typed in, for example, 5, then 7 (for 57), then a space bar, then the multiplication key, then 4, then 3. The output typewriter then produced an equals sign then the answer, "2451." The input typewriter then advanced a line, ready for another calculation.

Despite the fact that the machine had several obvious applications in business, Torres-Quevedo made no plans to produce it commercially.

In his writings, Torres-Quevedo declared his admiration for the research of Charles Babbage (see page 40), and laid out a vision for something not dissimilar to the robots we see today. "A special chapter of the theory of machines which would be called automatics. The means of constructing automata endowed with a pattern of behavior of greater or lesser complexity should be investigated," he wrote. "These automata will have sense organs, i.e., thermometers, magnetic compasses, dynamometers, manometers etc, mechanisms for sensing the conditions which should have an influence on the operation of the automata."

Today, El Ajedrecista is on display at the Torres-Quevedo Museum of Engineering at the Universidad Politécnica de Madrid. It would take another seventy years before a machine could play a full game of chess (rather than an endgame where the machine player was in an unbeatable position) and beat the world's best player (see page 119).

1914

RESEARCHER:
Karel Čapek

SUBJECT AREA:
Robotics

CONCLUSION:
The play that introduced the word was far-sighted in other ways, too

WHAT DOES "ROBOT" MEAN?

HOW KAREL ČAPEK'S PLAY INVENTED THE WORD "ROBOT"

The word "robot" does not come from science, but from science fiction, much like the phrase "atomic bomb," which was coined by the writer H. G. Wells. Czech playwright Karel Čapek said he decided on the word "robot" after explaining the plot of a play he was writing to his brother, Josef, a painter.

The plot of what would become Čapek's science fiction play *R.U.R.* (Rossum's Universal Robots) sounds eerily familiar to those of Hollywood science fiction films released in the century after it came out: a genius scientist has a technical breakthrough allowing him to create thousands of synthetic slaves, then the slaves rebel against their masters and wipe out the human race.

"Call them roboti," said Josef when Karel had explained his idea. "Roboti" was a Czech word for laborer or serf. Čapek had previously thought of "labori," but preferred Josef's idea, thinking "labori" was too bookish. The idea stuck, and Karel wrote his play.

The play premiered at the National Theatre in Prague in 1921, in what was then Czechoslovakia. The play was a success not just in Czechoslovakia, but around the world, being a hit throughout Europe in the 1920s. During the 1930s, a version was produced for U.S. radio and another for BBC TV.

Not everyone was enamored of the play. Science fiction writer Isaac Asimov, who wrote dozens of novels about robots and devised the Three Laws of Robotics (see page 76), said, "Čapek's play is, in my own opinion, a terribly bad one—but it is immortal for that one word." But then Asimov's view of robots was highly optimistic: and Čapek's was not. Čapek's idea influenced a vast amount of subsequent science fiction, from the remorseless Terminator that always comes back to life, to the androids who revolt against human rule in Blade Runner.

Synthetic Servants

In Čapek's play, robots are made of engineered chemical flesh, not metal, and are manufactured by the thousands in a huge factory, looking identical to people but built to be the slaves of human beings. One grade of robot is built for manual labor and is manufactured in "vast quantities."

In the play, the factory manager Harry Domin says that the manual labor robots are "as powerful as a small tractor. Guaranteed to have average intelligence." The robots are deliberately robbed of their ability to be creative, or to have emotions, to make them more successful as workers. A shocking scene in Act One has one of the factory proprietors offer to dismember a humanlike secretary robot to prove that she is not "real."

The play established not just the word "robot" but the idea of artificial humans, long before there was the technology to deliver a machine remotely resembling a human being. Čapek wasn't imagining a specific technology being used to create the robots, but instead offers a parable about the dangers of technology and human greed.

The play introduced many of the stock themes that have come to dominate our image of robots and AI—specifically the idea of robots as a threat.

Rise of the Robots

At the end of the play, the robots explain their justification for wiping out the human race. They were inspired, of course, by the vile acts of human beings themselves. When the last survivor of the human race asks why they have killed every single human being, one of the robots replies, "We wanted to be like people. We wanted to become people."

1925

RESEARCHER:
Francis P. Houdina

SUBJECT AREA:
Self-driving cars

CONCLUSION:
"Phantom autos" took to the streets in the 1920s

CAN A ROBOT DRIVE ITSELF?

HOW THE HOUDINA "AMERICAN WONDER" SPARKED SELF-DRIVING CARS

The first driverless cars weren't referred to as "self-driving" or "autonomous," they were known as "phantom autos." The cars appeared almost a century before self-driving car technology would be a commercial reality, and were used to educate people about road safety, which was deeply ironic given the safety of the "phantom" cars themselves.

Taking to the streets in the 1920s and beyond, phantom autos were fully remote controlled, and were piloted via radio from another car (and in one case, from a plane flying overhead).

One, the Houdina car, described as the "American Wonder," made a significant splash (not to mention a crash) on its debut in New York in 1925. Long before the dawn of health and safety regulations, the self-driving car was shown off in busy streets full of people and traffic.

"The Man Did Not Approach the Controls"

There had been demonstrations of driverless vehicles before, including a remote-controlled tricycle shown off by Leonardo Torres-Quevedo in 1904, but this was a full-sized production car, driven through busy city streets without anyone at the helm.

Time Magazine wrote, "In Manhattan, an empty touring car lounged against a Broadway curb. A man stepped on the running-board but did not approach the controls. Pedestrians gaped to hear the chauffeurless machine start its motor, shift into gear, lurch away from the curb into thick traffic."

The car was piloted by "Francis P. Houdina," said to be a pseudonym used by two young engineers. It drew crowds, but not everything went entirely to plan. In the second car, John Alexander of the Houdina company, sent a signal to the receiver in the first car, but the driving apparatus attached to the shaft in the other car failed due to a loose housing. At that point, everything became quite alarming.

"The radio car careered from left to right, down Broadway, around Columbus Circle and south on Fifth Avenue, almost running down two trucks and a milk wagon," wrote *The New York Times*. "At 47th St., Houdina lunged for the steering wheel but could not prevent the car from crashing into the fender of an automobile filled with camera men." The police at this point implored Houdina to stop his experiment, but he continued back onto Broadway and Central Park Drive.

The car was a fairly simple device: a Chandler sedan rigged with a radio antenna and small electric motors that controlled the vehicle's speed and steering, "driven" by a crew of engineers trailing closely behind the vehicles.

The "receiver" was a kite-shaped radio antenna. The car also had some form of belt attached to the steering column, plus devices to start the car, accelerate, and brake—although it's unclear whether the engineers also had control over the clutch and gear shift.

No Escape

The Houdina car also attracted some unwanted attention from the magician and escape artist Harry Houdini, who was so enraged by the similarity between the names Houdina and Houdini that he went to the offices of the Houdina Company. Finding a package addressed to "Houdini" in their offices, he went berserk. As *The New York Times* reported, "he tore from a packing case a tag addressed 'Houdini' [...] and when ordered to return it refused, seized a chair and broke an electric chandelier when they tried to prevent him from leaving the room." The company denied that the name "Houdina" was intended to impersonate Houdini.

Safety First

Houdina's invention inspired a wave of such phantom autos and "phantom cars," which were used in advertising and in demonstrations in small towns across the United States. Some were piloted by radio remote control, others via wired connection between cars, and on at least one occasion, from a low-flying aircraft, where a car performed U-turns in the street, controlled from overhead.

Ironically, given the outcome of Houdina's first test, phantom cars were also used in pioneering road safety campaigns. The roads of the 1920s were far more dangerous than they are today (due to the lack of driver training and road safety measures), and phantom auto drivers hoped to inspire their human counterparts to take more care.

Phantom audio operator J. J. Lynch spoke to the *Daily Times-News* of Burlington, North Carolina, in 1937, saying, "Regular safety lectures leave a sour taste in everybody's mouth, especially when you start telling another fellow about his shortcomings as a driver. But when you give them this kind of demonstration and talk about safety at the same time, they listen to you."

Publicity to Practicality

In later decades, interest in the idea of self-driving vehicles continued quietly. At the Futurama exhibit in 1939's World Fair, General Motors worked with designer Norman Bel Geddes to unveil a vision of the future, with cars driving vast highways under "automatic radio control." In Britain in 1963, a Citroën zoomed along at up to 80 miles per hour (130 kph) with Conservative politician Lord Hailsham sitting in the front, with his hands off the wheel, guided by wires under a specially prepared test track.

But it would take almost half a century before self-driving technology would be practical (and safe), with a technology gold rush kick-started by a hair-raising driverless race in the California desert, the DARPA Grand Challenge (see page 142).

CAN A ROBOT RESPOND TO INSTRUCTIONS?

HOW HERBERT TELEVOX DID A HUMAN JOB

1927
RESEARCHER:
Roy J. Wensley
SUBJECT AREA:
Humanoid robots
CONCLUSION:
The first humanoid robot that could perform a useful task

In the 1920s, several "mechanical men" dazzled audiences around the world, with metal bodies and futuristic looks. Most were simply automata, animated by trickery, as had been seen in previous centuries (see page 14)—but with the addition of twentieth-century technologies such as electricity and compressed air.

But one, the Televox, shown off by Westinghouse in 1927, could do useful work, despite the humanoid machine arriving more than seven decades before a robot could walk like a human being (see ASIMO, page 127).

The Televox—or to give him his full name, Herbert Televox—was capable of receiving audio instructions in the form of audible tones via phone, and activating machinery in response. It looked very much like a box full of electrical machinery attached to a cut-out figure of a human being (and in fact, that's basically what it was).

At the worldwide launch of Herbert Televox, its creator also revealed that in the company's laboratory, it had built a door that would open in response to the words "open sesame," but said that voice recognition was too unreliable for use via the phone system, so Herbert Televox communicated solely via buzzing and chirps sent via the phone system. The hubbub around Televox was a tribute to the marketing genius of Westinghouse's Roy J. Wensley.

Machines that Think?

A breathless editorial in Popular Science Monthly was headlined, "Machines That Think" and described the Televox's powers in gushing terms: "Electrical men answer phones, do household chores, operate machinery, and solve mathematical problems." *The Manchester Guardian* took a slightly more pragmatic view, with a piece headlined, "Starting the oven by phone."

In a demonstration in New York, Wensley showed off Herbert Televox responding to signals (generated with a tuning fork) and pressing the appropriate switch in response. Herbert Powell of *Popular Science Monthly* wrote, "The mechanical man is not connected electrically to the telephone, but listens much as you would. His ear is a sensitive microphone placed close to the receiver. His voice is a loudspeaker close to the transmitter. And the language he speaks is a series of mechanically operated signal buzzes."

The Televox would be the first in a series of increasingly large robots made by Westinghouse, which at the time was pioneering new ideas about how to cut down the number of human staff at remote electricity substations.

Pitch Perfect

A Televox unit in a substation could receive commands of a certain pitch (created by tuning-fork oscillators), process the codes, and respond by, for example, opening a switch.

The robot communicated with a second Televox unit in a command center, making coded noises to indicate that the commands had been carried out.

Westinghouse also showed off how the system could check the water level in a reservoir, by communicating with Televox apparatus connected to a water level meter, which gave a set number of buzzes to indicate how high or low a water level was. The apparatus was in use in New York by 1927.

The Manchester Guardian wrote, "Mr. Wensley, explaining this system, said that sounds that come over the telephone to the televocal apparatus are received from the receiver by a sensitive microphone, and the buzzing signals made by it are given out by a loudspeaker close to the telephone transmitter. When the bell rings, a sound-sensitive relay lifts the telephone hook, starts up the station-signal buzzer, and sets the whole apparatus ready for action."

Much of the hype around Televox was fairly fanciful: the unit did not do household chores (although in theory it could), nor was it particularly good at math. But Wensley's decision to make it look like a human being (and use it in advertising and press appearances) made the Televox a sensation in America and Europe.

A Very Fine Brain

It would be the first of several robots made by Westinghouse, culminating in Elektro, a voice-controlled robot that introduced itself at the 1939 World's Fair with the words, "Ladies and gentlemen, I'll be very glad to tell my story. I am a smart fellow as I have a very fine brain of forty-eight electrical relays."

At the Westinghouse Pavilion, the robot stood high above the audience on a platform and even "walked" (albeit with an odd sliding movement). The robot used a record player to simulate conversation and had a "vocabulary" of 700 words, and could smoke cigarettes and blow up balloons. The next year, he reappeared with his own metal dog, Sparko. Elektro had cost hundreds of thousands of dollars to develop, and went on tour where he was seen by millions of people.

He was not billed as a "robot" as the word hadn't quite gained the currency it has today. Instead, he was described as a "Moto-man." Elektro's story had an unexpected epilogue when, long after retirement, the robot was drafted in to play the part of the robot Thinko in the racy 1960 movie comedy *Sex Kittens Go to College*.

Today, the surviving pieces of Herbert Televox and Elektro are on display in Ohio's Mansfield Museum.

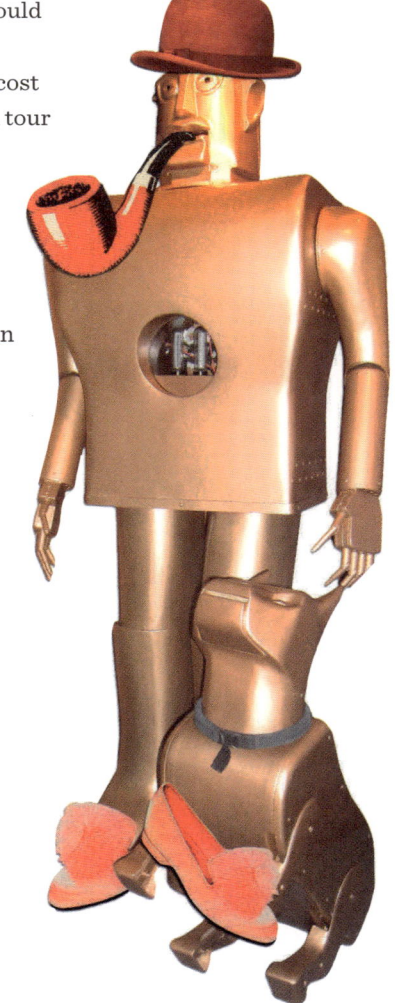

1928

RESEARCHER:
Fritz Lang

SUBJECT AREA:
Robots in fiction

CONCLUSION:
The "Maschinenmensch" inspired the look of robots in fiction and reality

WHAT SHOULD "MAN MACHINES" LOOK LIKE?

FROM MOVIES TO REALITY

Few movie costumes have become as iconic as the strange metal woman in Fritz Lang's 1927 silent movie masterpiece *Metropolis*—known as the Maschinenmensch or "man-machine." Appearing in one iconic scene on a thronelike seat surrounded by lights, the man-machine has a blank, unsettling metal mask and a metallic female body, surrounded by strips of metal that give the creature a look of industrial machinery.

The creature's body is clearly female, its movements jerky, exaggerated, and machinelike—and the design's influence on later robots in fiction (and reality) raises important questions about the sexualization of technology.

The costume (which famously went missing during the shoot) was inspired, in part, by the mask of the boy-king Pharaoh Tutankhamun, which had been unearthed in the Valley of the Kings just a few years earlier in 1922. The mask was built by designer Walter Schulze-Mittendorff.

The robot costume was constructed on top of a plaster cast of the actress Brigitte Helm, who played both the Maschinenmensch and her human double: chaste, oppressed worker, Maria. It used what Schulze-Mittendorff described as "plastic wood—a pliable wooden material, fast-hardening on exposure to air, with the possibility of being able to work on it like naturally grown wood."

Behind the Mask

The end result was human but inhuman, with a blank metallic face, and long panels over its limbs. The creature is introduced in a scene where its mad scientist maker is ecstatic over how he will turn his metal creation into a real woman. Teenage actress Brigitte

Helm wore the costume throughout the film, in long scenes (the perfectionist director Lang filmed hundreds of hours of footage).

Helm's mother had sent a picture of her daughter to director Fritz Lang's wife Thea von Harbou (who wrote the novel that became *Metropolis*), and the director gave the completely unknown actress the female lead role in the film. She was just sixteen when she took her screen test.

Helm played both Maria and the sexualized robot that impersonates her. The costume she wore was rigid and uncomfortable, with the cast of her body having been taken while standing up. The result was that after filming, she was covered in cuts and bruises.

At one point Helm asked why a double could not take her place in one arduous scene, which took nine days to film, where her face didn't appear in shot. Lang said, "I have to feel that you are inside the robot. I was able to see you even when I didn't."

The image of the man-machine went on to inspire future film robots including *Star Wars's* C-3PO, which was based largely on the look of Lang's Maschinenmensch, and who, in turn, inspired roboticists such as Cynthia Breazeal to create "social robots" to interact with human beings. The look of Lang's fictional robot also inspired the designers of real robots. Hajime Sorayama, who designed the look of Sony's iconic robot dog Aibo (see page 124), showed off a huge sculpture inspired by the Maschinenmensch in 2019.

Feet of Clay

Lang's film has become an enduring icon of the black-and-white era, but it was initially a flop and nearly bankrupted the German studio, UFA, which made it. At the time, it was the most expensive film ever made, budgeted at around seven million Reichsmark, and on launch was a disaster, hated by both critics and the public. *The New York Times* described the film as a "technical marvel with feet of clay."

Metropolis depicts the year 2006, when a ruling class lives at the top of skyscrapers while the working class toil beneath them in slavelike conditions. The Maschinenmensch is devised by the scientist Rotwang and built at the behest of an authoritarian ruler. The robot Maria turns into a real woman (due to technology, we're told, but in reality probably due to budgeting constraints) and attempts to sow discord among the workers in a bleak futuristic city.

The false Maria tells the workers, "Who is the living food for the machines in Metropolis? Who lubricates the machine joints with their own blood? Who feeds the machines with their own flesh? Let the machines starve, you fools! Let them die!" The robot's evil actions lead to it being burnt at the stake, returning to its true metallic form.

The Woman Machine

Like many of the robots that would follow in fiction, it's a disturbing, frightening character. The robot is also highly sexualized, and prefigured other problematic fictional (and real) robots and AI helpers. Many of the "gynoid" or female robots inspired by the Maschinenmensch have been highly sexual, and portrayed as being created by men to do their bidding.

In later films, characters such as the android sex worker Pris in *Blade Runner* (who goes rogue and has to be violently "retired") and the submissive female robots of the 1975 film *The Stepford Wives* (based on Ira Levin's satirical novel where successful men manufacture compliant wives) are sexual and slavelike. They beg troubling questions including why so many real "servant" robots have taken female form. Even now, the default voices for helpful "voice assistants" such as Alexa and Siri are often female.

Lang went into exile in America as the Nazi party rose to power, whereas von Harbou made films for the Nazis, being detained by British authorities at the end of the war. After making the film, Brigitte Helm went on to have a successful career at UFA—but refused to work with Lang again.

WHAT USE WAS POLLARD'S PATENT?

WHY A "POSITION-CONTROLLING APPARATUS" PAVED THE WAY FOR ROBOT ARMS

1938

RESEARCHER:
Willard Pollard
SUBJECT AREA:
Robot arms
CONCLUSION:
Used a "pantograph" to create a design for a spray-painting robot

In fiction, robots tend to look like human beings: but in the workplace, robots have tended to take the form of disembodied robot arms, from limbs that perform surgery to bomb-disposal arms mounted on carts, to the famous Canadarm used on the Space Shuttle program, which captured satellites and positioned astronauts.

The first designs for robot arms actually appeared before World War II (although it would be some time before the full business potential of the idea would be realized).

In 1938, American engineer Willard Pollard filed a patent for what was described as a "position-controlling apparatus" (a robot arm). It was a machine that he hoped would be used in America's car industry to automate the spray-painting process.

In the first half of the twentieth century, America led the world's car industries, largely due to streamlining and automating processes that had previously been done by hand.

When Henry Ford installed the first moving assembly line for mass-production of cars, it reduced the time to build a car from more than twelve hours (when done by teams of men working together on one vehicle) to just over an hour and a half with vehicles moving between teams of workers down a production line. The knock-on effect on the price of the vehicles enabled Ford's Model T to dominate the industry, selling 10 million in the next decade.

Pollard's proposed innovation was the next logical step: an automated, programmable arm that could take over one of the jobs on the way: spray-painting the car. There were two related patents, one for an automatic control system for spraying the car, and one for the arm itself, filed in 1934 and 1938. The patent for the

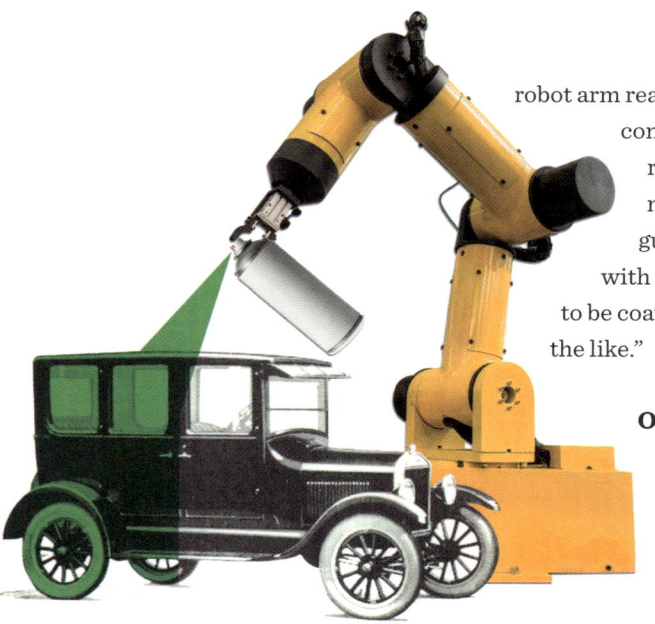

robot arm read, "My invention relates to position-controlling apparatus. More specifically it relates to apparatus for controlling the movement and positioning of a spray gun for controlling it in its movement with respect to curved or irregular surfaces to be coated, such as an automobile body or the like."

Out on a Limb

Willard's invention wasn't, of course, a complete leap in the dark. The machine described in his patent was a pantograph, similar to devices used to make multiple copies of writing, by linking one pen to another (or several others) with the pens held on an "arm" articulated by a series of joints. Pantographs were first described by the Greek philosopher and mathematician Hero of Alexandria (himself a deft maker and designer of automatons, see page 14).

Pantographic arms had also been used in the notorious fake robot the Mechanical Turk, a supposed mechanical chess player that defrauded audiences in the late eighteenth century. A hidden human chess player inside the Turk had moved one pantographic arm, and in response, the supposedly "automated" Turk would move its robotic arm to place a piece on the chess board.

Man in the Machine

But what was unique about Pollard's arm is that it did not require a man inside the machine—or indeed any human control at all. The arm had "five degrees of freedom" (referring to the different ways it could move, such as rolling, pitching (up or down), and yawing (left and right).

What made it useful was that it could be reprogrammed to paint in different patterns, by changing a strip with instructions, or even fitted with different tools for a different part of the assembly line process.

The machine used pneumatic cylinders to control its position, and was "programmable" (in a rudimentary sense) so that it could be rapidly switched from one job to another. Pollard wrote, "If it is desired to paint a 'coupe' coming up the line, a record '43' will be selected, which has a record suitable for this model; if a 'sedan,' another record will be selected; and so on."

Right to Bear Arms
But Pollard's idea was ahead of its time. The pantographic arm never went into mass production, although some have suggested that the DeVilbiss paint company may have built a prototype robot arm in the early 1940s, based around Pollard's design or a related patent filed by Harold Roselund's "Means for Moving Spray Guns or Other Devices Through Predetermined Paths."

In his article "Evolution of Robotic Arms," Michael Moran notes, "The modern era of robotics was launched by the intrepid use of these two, little known arms developed in the late 1930s." Moran says of Pollard, "his design and interest in an industrial application for automated robotic arms would spur on the ingenuity of others."

The advent of World War II would see a leap forward in sensors and computing, and a move toward systems that could be used to control "intelligent" machinery. It would take two more decades before Pollard's dream of robots invading American car factories would become reality, with the world-changing invention of the Unimate robot arm (see page 95).

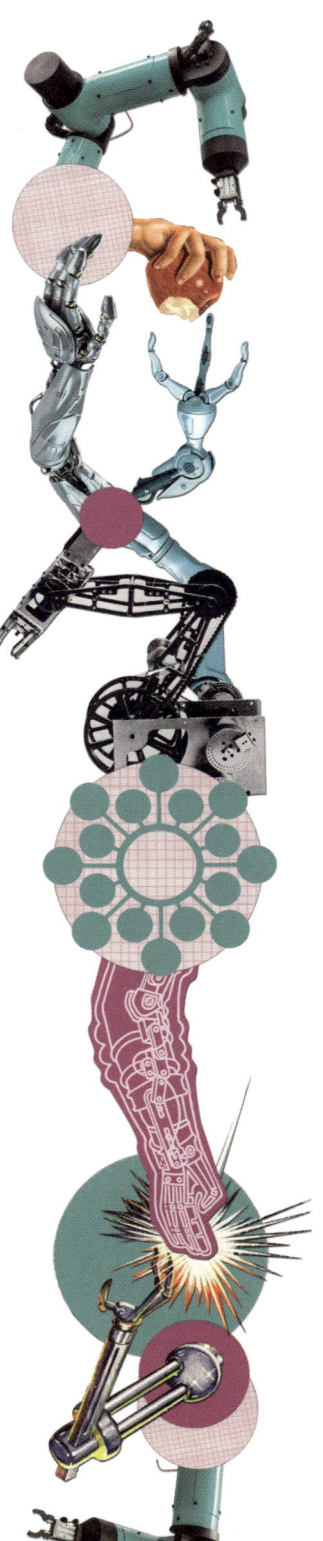

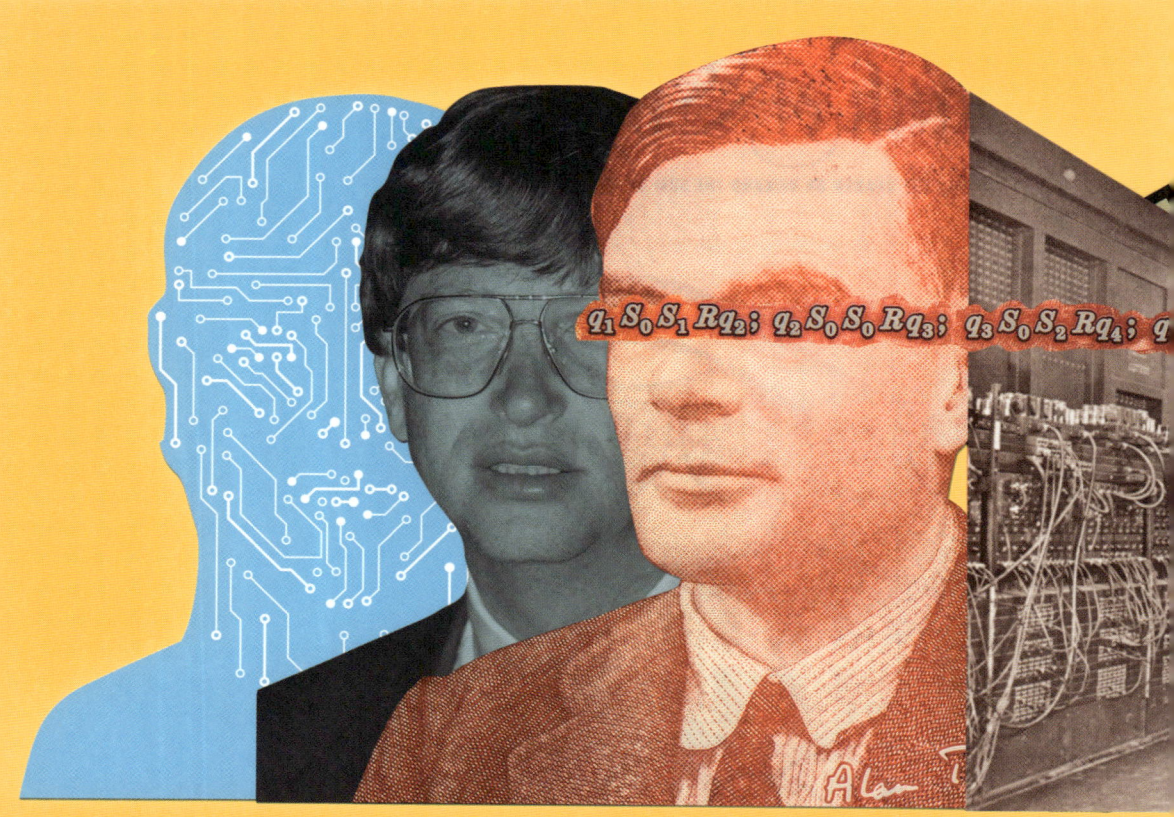

CHAPTER 4: Developing Intelligence 1940–1969

The second half of the twentieth century saw technology born out of World War II offer a boost to the emerging fields of computing and artificial intelligence, with highly classified machines such as the ENIAC built in the dying days of the war offering a glimpse of the capabilities of programmable general-purpose computers.

There was also an explosion in ideas, with wartime computing pioneer Alan Turing suggesting a test that could pinpoint whether a machine is truly intelligent, and delegates at a conference in Dartmouth, New Hampshire, in 1956

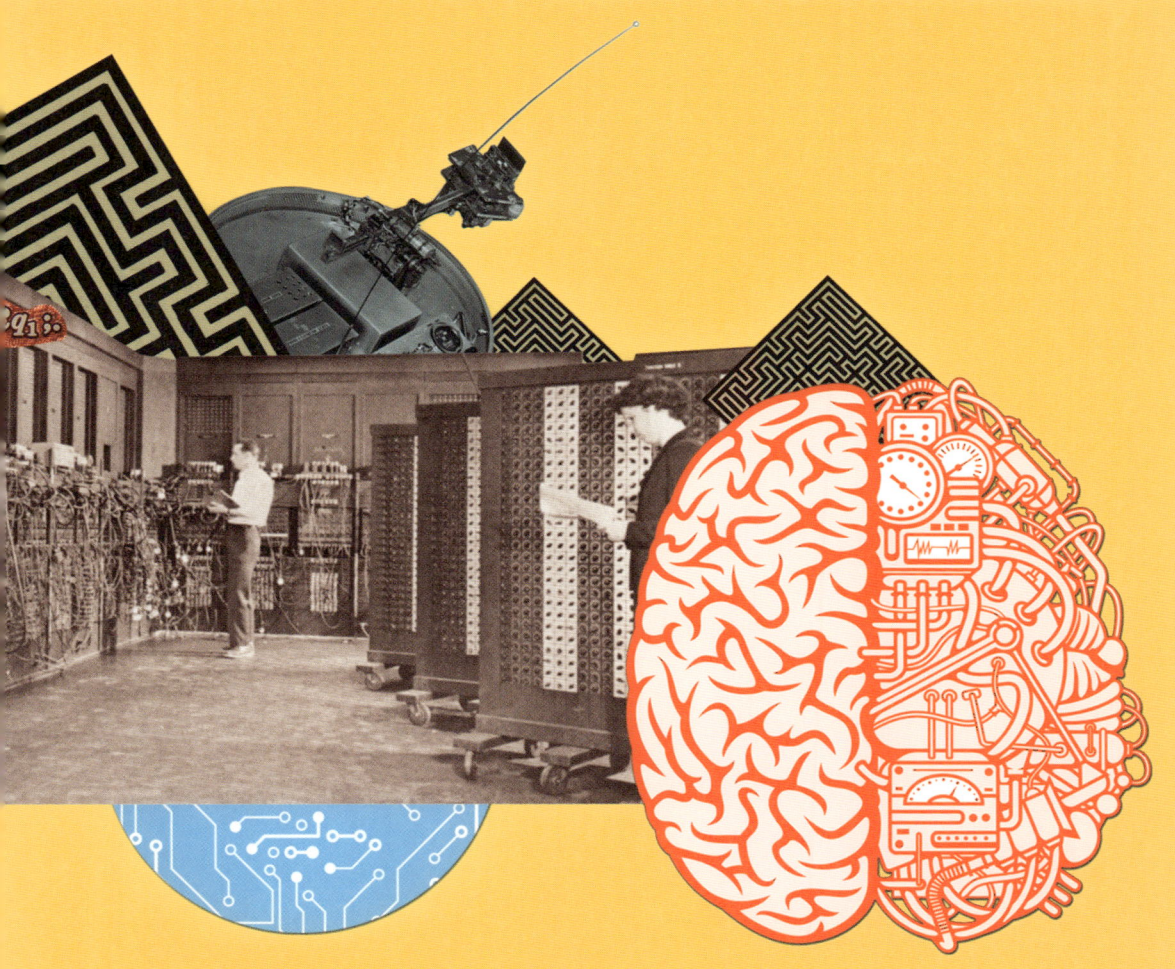

coining the term "artificial intelligence," and offering some very optimistic ideas about when true machine intelligence might arrive.

But while AI research was heading for what is described as the "artificial intelligence winter," some intriguing robots were taking shape in laboratories across the globe, with the aptly named "Shakey" able to navigate through mazes under its own steam, and inspiring people around the world, including a young Bill Gates.

1942

RESEARCHER:
Isaac Asimov

SUBJECT AREA:
Robot behavior

CONCLUSION:
Created "laws" to ensure robots don't harm human beings

ARE ROBOTS ABOVE THE LAW?

HOW ASIMOV'S "LAWS OF ROBOTICS" HELP US IMAGINE A HUMAN-ROBOT SOCIETY

The word "robotics" was coined by the prolific science fiction writer Isaac Asimov, who said he was completely unaware that he was creating a new word, and had just assumed that it had already existed before he came up with it in the early 1940s.

Asimov's "Three Laws of Robotics" come from his *Robot* science fiction novels, and are among the most famous ideas Asimov conceived, and the subject of serious debate even today.

The Three Laws of Robotics are a simple series of rules, designed to ensure that robots are helpful servants to human beings (and don't turn on their masters).

The Three Laws are: "A robot may not injure a human being or, through inaction, allow a human being to come to harm.

"A robot must obey the orders given to it by human beings except where such orders would conflict with the First Law.

"A robot must protect its own existence as long as such protection does not conflict with the First or Second Laws."

In his later fiction, Asimov added a fourth law (or as he termed it, the "zero-th" law): "A robot may not harm humanity, or, by inaction, allow humanity to come to harm."

A prolific writer, Asimov wrote staggering volumes of books in his lifetime (including everything from detective fiction starring himself to guides to Shakespeare), saying that he started writing every day at 7:30 a.m. and finished at 10 p.m.

He was known for his huge muttonchop whiskers and claimed that he would only ever rewrite his work once, so he could keep up his famously frenzied rate of production.

He said, "I make no effort to write poetically or in a high literary style. I try only to write clearly and I have the very good fortune to think clearly so that the writing comes out as I think, in satisfactory shape."

Asimov refused to fly, and only took exercise on a cross-country skiing machine in his apartment. He had been born in Smolensk, and credited his work ethic to his Russian father, who owned sweet shops in Brooklyn that were open 6 a.m. to 1 a.m. seven days a week, and in which Asimov worked daily from a young age.

He said, "I do all my own typing, my own research, answer my own mail. I don't even have a literary agent. This way there are no arguments, no instructions, no misunderstandings. I work every day. Sunday is my best day: no mail, no telephones. Writing is my only interest. Even speaking is an interruption."

An Influential Example

Many roboticists featured in this book say they were inspired by Asimov's work, from George Devol and Joseph Engelberger, creators of the first industrial robotic arm (see page 95) to Yoshiyuki Sankai, CEO of Cyberdyne (see page 145). Other famous Asimov fans include Amazon's Jeff Bezos.

Asimov wrote thirty-seven novels and short stories in his *Robot* series, imagining a future where helpful "positronic robots" lived alongside humanity, guided by the three laws. His first *Robot* stories were published in magazines in the 1940s and compiled as *I, Robot*, in 1950.

Asimov's robots were unlike the unfeeling machines that rose up against humanity in Karel Čapek's play *R.U.R.*, being portrayed as benevolent, in forms such as a child's robot nursemaid Robbie to a steadfast robot policeman.

In "Robbie," a short story, George Weston protests after his wife wants to get rid of their robot nursemaid (as the child has become too attached to it). He says, "A robot is infinitely more to be trusted than a human nursemaid. Robbie was constructed for one purpose really—to be the companion of a little child. His entire mentality has been created for the purpose. He just can't help being faithful and loving and kind.... That's more than you can say for humans."

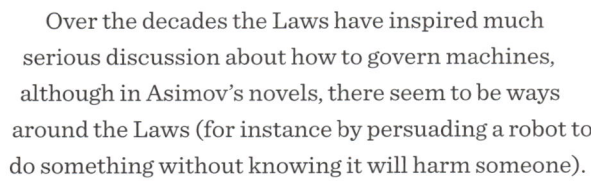

Over the decades the Laws have inspired much serious discussion about how to govern machines, although in Asimov's novels, there seem to be ways around the Laws (for instance by persuading a robot to do something without knowing it will harm someone). The tag line of the 2004 Will Smith film based on Asimov's *I, Robot* was simply: "Rules are made to be broken," with the film's plot involving a murder committed by a robot.

One Law for Them

Asimov's Laws are still the basic jumping-off point for many discussions of robots and ethics in artificial intelligence. But over the years, roboticists have pointed to various problems with the laws, including the fact that robots are allowed to intentionally harm other robots.

In Britain, the Engineering and Physical Sciences Research Council attempted to come up with improved laws for robotics, noting, "Asimov's rules are fictional devices. They were not written to be used in real life and it would not be practical to do so, not least because they simply don't work in practice.

"For example, how can a robot know all the possible ways a human might come to harm? How can a robot understand and obey all human orders, when even people get confused about what instructions mean?"

The proposed new Laws included a prohibition on robots designed to kill people (something that many campaigners view as an increasingly serious threat, see page 130).

Another proposed Law makes designers and builders of robots responsible for the acts of their creations. The Law says, "Humans, not robots, are responsible agents. Robots should be designed and operated to comply with existing law, including privacy."

HOW DID WOMEN HELP ENIAC?

THE HARD-WORKING THINKING MACHINE

1944

RESEARCHERS:
John Mauchly, Frances Holberton
SUBJECT AREA:
Digital computing
CONCLUSION:
Computers can outpace humans at many tasks

By the time the room-sized ENIAC (Electronic Numerical Integrator and Calculator) "retired" from duty in 1955, it was estimated that it had done more arithmetic in its decade-long working life than the entire human race had done in all the centuries beforehand. The 24-ton, 1,800-square-foot (167 m²) machine was made up of vacuum tubes and diodes, and had been commissioned during World War II, with construction beginning in 1943.

Calculating Gun Trajectories
ENIAC was a result of a proposal by scientist John Mauchly, who put forward the idea of a vacuum-tube-based machine to speed up calculations used by the U.S. military. It would be the first programmable general-purpose electronic digital computer.

The ENIAC was built to solve the problem of creating firing tables, used to calculate trajectories for guns under standard conditions. During the war, a huge number of these tables were required by the American military for new weapons under development.

Initially, the trajectories were calculated manually with mechanical calculators, and a single sixty-second trajectory could take up to twenty hours. It proved so time-consuming that at one point the Ballistic Research Laboratory of the U.S. Army had more than 100 female students working just on trajectories.

In contrast, ENIAC could perform the same trajectory calculation in just thirty seconds. It could do 5,000 additions per second, or 360 multiplications of two ten-digit numbers. It also had settings for division and square roots. It was by far the most complex electronic system built to that point, with 17,000 vacuum tubes, 70,000 resistors, and 1,500 mechanical relays.

It sat in a 50-foot by 30-foot (15 x 9 m) basement in the University of Pennsylvania, and generated 174 kilowatts of heat as it operated, requiring its own air conditioning system. Its original cost estimate was $150,000, but this rose to $400,000 during production.

ENIAC, however, would never see active service, being completed only in November 1945, months after the end of the war. It would, though, go on to work on the development of America's hydrogen bomb.

Working the Machine

A shortage of male engineers as a result of World War II meant that many young women were drafted in to work on the ENIAC. The young women programmers, many of them mathematics graduates, were tasked with "hard-wired" programming, meaning that they spent much of their time setting switches and cables inside the machine itself. Many had first been hired to do calculations by hand, before working on the machine that, ironically, was built to take over their jobs. In fact, they themselves had previously been known as "computers."

Setting up ENIAC for new calculations took several days, with the programmers attaching wires to plugboards, then taking further hours to test that the machine had been configured correctly. Frances Holberton, who would go on to work on the computer languages COBOL and FORTRAN, was among the most intuitive at working out the correct paths through the machine. She said that ideas frequently came to her in her sleep.

Turning off the ENIAC was frowned upon, as switching it on and off tended to blow out the vacuum tubes needed to make calculations. The vacuum tubes blew out regularly, requiring staff to dig around and find out which one had blown and then replace it, a process that the team managed to refine down to just fifteen minutes.

The H Bomb

After the war, ENIAC's first real job was calculations relating to America's hydrogen bomb program, then in its infancy at Los Alamos. Later, Nicholas Metropolis of the Manhattan Project designed a new computer specifically for the hydrogen bomb, the superbly named MANIAC (Mathematical Analyzer, Numerical Integrator and Calculator), while ENIAC was eventually retired after it was said to have been damaged by a lightning strike.

In an illustration of just how far computing has come since the dawn of the computer age, students at Moore University commemorated the 50th anniversary of ENIAC in 1997 by creating a simulation of the entire machine on just one computer chip, with the hand-connected wiring of the original ENIAC emulated, and controllable from a PC.

The ENIAC itself was divided up, with some panels going to the Smithsonian and some in the University of Michigan. In later years, some panels were bought by billionaire Ross Perot and are now on display in the Fort Sill Field Artillery Museum in Oklahoma.

1949

RESEARCHER:
Edmund Berkeley

SUBJECT AREA:
Intelligent machines

CONCLUSION:
Helped to usher in the era of personal computing

CAN MACHINES THINK LIKE US?

HOW "GIANT BRAINS" HELPED US IMAGINE A COMPUTER IN EVERY HOME

In the early days of computing, it was common for people to think of the machines not so much as electronic tools, but as something more similar to human beings (or at least our brains).

It was an image often used in news reports of early computers such as the ENIAC (see page 77), and one that was energetically proposed in the first popular book on electronic computers: *Giant Brains or Machines That Think*, by Edmund Berkeley, which painted a hopeful picture of a world transformed by such "giant brains."

It was published in 1949. The previous year Norbert Wiener had generated a large amount of publicity with his *Cybernetics*, which discussed self-regulating mechanisms. Berkeley's book, though, painted a vivid picture of a computer-infused future, which sparked the public imagination.

Strange Giant Machines

"Recently there has been a good deal of news about strange giant machines that can handle information with vast speed and skill," Berkeley wrote. "These machines are similar to what a brain would be if it were made of hardware and wire instead of flesh and nerve."

Some of Berkeley's conclusions were optimistic, to say the least. He wrote, "A machine can handle information; it can calculate, conclude, and choose; it can perform reasonable operations with information. A machine, therefore, can think."

Born in 1909, Berkeley was an actuary and computer pioneer who had seen several of the huge computers of the decade in person, and the book describes several of the machines that existed at the time (and imagines a future where such machines change everything).

The book was a hit, and widely influential. Patrick McGovern, the creator of the *For Dummies* series of computer books, was inspired to build a computer that could beat anyone at tic-tac-toe after reading *Giant Brains*. As a result, he won a scholarship to MIT.

Fear of Robots

The book also raised fears that would become a familiar part of discussions about robots and artificial intelligence—including the fear of widespread unemployment. John E. Pfeiffer wrote in *The New York Times*, "An important chapter discusses the social impact of large-scale computing machines. In the past, technological unemployment has been largely confined to people who work with their hands, but many white-collar workers may find themselves replaced by ensembles of vacuum tubes when commercial computers are manufactured in the hundreds." The article notes that as "less than a dozen" large-scale computers are in use, that this concern was one "for the future."

Berkeley also warned of a robot uprising, suggesting that in the future, robots might pose physical danger to human beings. A life-long campaigner against nuclear weapons, Berkeley wrote a fiery indictment of autonomous and technological weaponry.

Simple Simon

The book helped to spark public interest in computers, and also helped to ingrain the habit of referring to a computer as a "brain," with Berkeley predicting a future filled with such brains.

"Men have only just begun to construct mechanical brains," Berkeley wrote. "All those finished are children: they have all been born since 1940. Soon there will be much more remarkable giant brains."

Perhaps the most lasting legacy of the book is Simon, a simple "mechanical brain," which Berkeley built after describing it in the book. It's often referred to as the first personal computer.

Berkeley wrote, "Simon is so simple and so small in fact that it could be built to fill up less space than a grocery-store box; about four cubic feet. . . . It may seem that a simple model of a mechanical brain like Simon is of no great practical use. On the

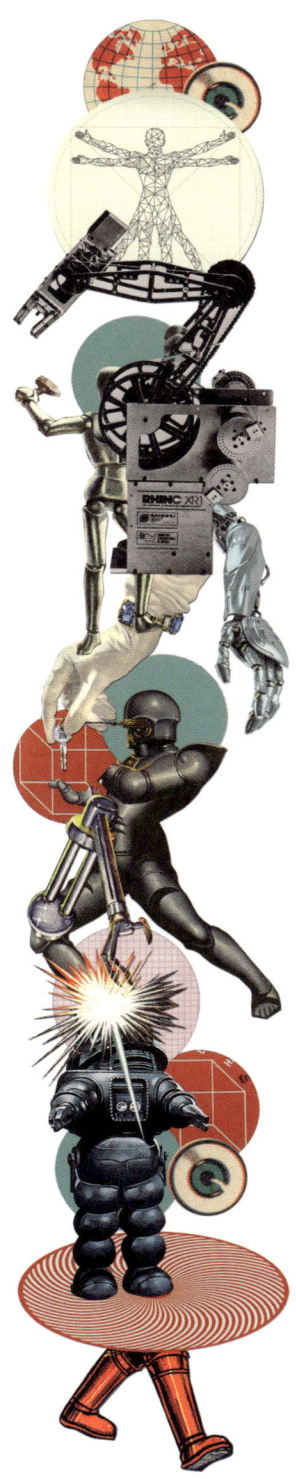

contrary, Simon has the same use in instruction as a set of simple chemical experiments has: to stimulate thinking and understanding, and to produce training and skill. A training course on mechanical brains could very well include the construction of a simple model mechanical brain, as an exercise."

Simon could perform simple sums, with data input via punched cards (punched-card machines were something Berkeley had worked with as an actuary). It gave its "answers" via lights at the back of the unit.

Modern Predictions

Berkeley hoped that the machine might spark a boom in building "mechanical brains"—similar to the crystal radio boom of the 1960s. The limitations of the machine (it could only show the numbers 0, 1, 2 and 3) meant that it never came to pass.

But his experience of the machine led him to make a famous (and fairly accurate) prediction of our modern world. In an article in *Scientific American* in 1950, Berkeley wrote, "Some day we may even have small computers in our homes, drawing their energy from electric-power lines like refrigerators or radios [...] They may recall facts for us that we would have trouble remembering. They may calculate accounts and income taxes. Schoolboys with homework may seek their help."

HOW CAN A MACHINE PASS THE TURING TEST?

ASSESSING A MACHINE'S ABILITY TO SHOW INTELLIGENT BEHAVIOR

1950
RESEARCHER:
Alan Turing
SUBJECT AREA:
Intelligent machines
CONCLUSION:
Artificial intelligence can impersonate people

How do you tell if a machine is intelligent? British computing pioneer Alan Turing—often described as the "father of artificial intelligence"—came up with a simple test in 1950, which he described as the "imitation game." In succeeding decades, it has become known as "the Turing Test." It's a simple parlor game, where a judge sits separately from two people: one a human, one a machine. The judge converses with them, and has to guess which one is human.

In his scientific paper "Computing Machinery and Intelligence," Turing suggests that a machine would have "won" the game if it could convince the judge it was human, although the rules have been interpreted several different ways in the decades since.

The Imitation Game

Turing presented two sorts of game, one with a man and a woman, where they have to fool the judge as to what sex they are, and one with a machine and a human. If a computer can fool the judge as to its humanity just as a human could fool the judge as to their sex, Turing suggests it would have won.

Turing acknowledges that the test simplifies the issue. He dismisses the question, "Can machines think?" and asks instead, "Are there imaginable digital computers that would do well in the imitation game?"

The test is not built to root out "true" intelligence, or even to understand it, but simply

to test whether a machine can imitate a human. He suggests that the machine should, essentially, "lie," in an attempt to fool its interrogator. He advises that a machine should pause for thirty seconds before answering complex maths questions, better to simulate a human contestant.

"I do not wish to give the impression that I think there is no mystery about consciousness," Turing wrote, "but I do not think these mysteries necessarily need to be solved before we can answer the question with which we are concerned."

Machines that Think

Turing himself was a little over-optimistic about how long it would take to create an intelligent machine. He predicted that, by the end of the twentieth century, machines would be able to "think." "The use of words and general educated opinion will have altered so much that one will be able to speak of machines thinking without expecting to be contradicted," he wrote.

In the more than half-century since Turing posed the question, AI chatbots have vied to "pass" the test in various different forms. Some researchers have even claimed "victory" in various Turing Tests, although none without controversy.

The first software capable of attempting the Turing Test was ELIZA, developed at MIT in the 1960s, which mimicked human conversation. Using "pattern matching" (looking for phrases and then replying with variants on the same phrase), it attempted to maintain a humanlike conversation. But creator Joseph Weizenbaum believed ELIZA showed the flaws of Turing's Test, as ELIZA had no comprehension whatsoever of what people were saying to "her."

The annual Loebner Prize, launched in 1990 by inventor Hugh Loebner, saw chatbots compete to fool a panel of judges that they were human, with dozens of bots vying for prizes over the years.

Is Eugene Goostman Real?

In 2014, researchers claimed that a computer program named Eugene Goostman had passed the Turing Test at an event at the

Royal Society in London. The software was designed to simulate the conversation of a thirteen-year-old Ukrainian boy and was developed in St. Petersburg in Russia. University of Reading researcher Kevin Warwick claimed victory after Goostman fooled thirty-three percent of the judges in five-minute unrestricted conversations.

"This event involved the most simultaneous comparison tests than ever before, was independently verified and, crucially, the conversations were unrestricted," Warwick said. "A true Turing Test does not set the questions or topics prior to the conversations. We are therefore proud to declare that Alan Turing's test was passed for the first time."

Others were more skeptical, describing it all as a "PR stunt," and pointing out that various bots had previously achieved similar success. Warwick was himself no stranger to publicity-grabbing events, having previously implanted a computer chip in his arm and describing himself as "the first cyborg." Critics also suggested that the approach of the Goostman bot is unfair, as it uses its supposed youth and Ukrainian origins to mask mistakes, pretending these are caused by youth or cultural differences.

With corporations increasingly deploying chatbot software as their first line of communication with customers (and with many of us communicating with voice assistants such as Siri and Alexa), software similar to what Alan Turing imagined is all around us on a daily basis, conversing perfectly naturally. Crucially, though, these bots never try to fool us that they are human.

Scientists no longer consider the Turing Test to be a true benchmark of any kind of artificial intelligence, but it remains an important part of daily life. All of us experience a form of "reverse Turing Test" on a regular basis, in the form of CAPTCHA questions in online forms, designed to root out machines impersonating human beings. Every time you pick out images of palm trees or fire hydrants in an attempt to prove you are not a robot, you are doing Turing's test in reverse.

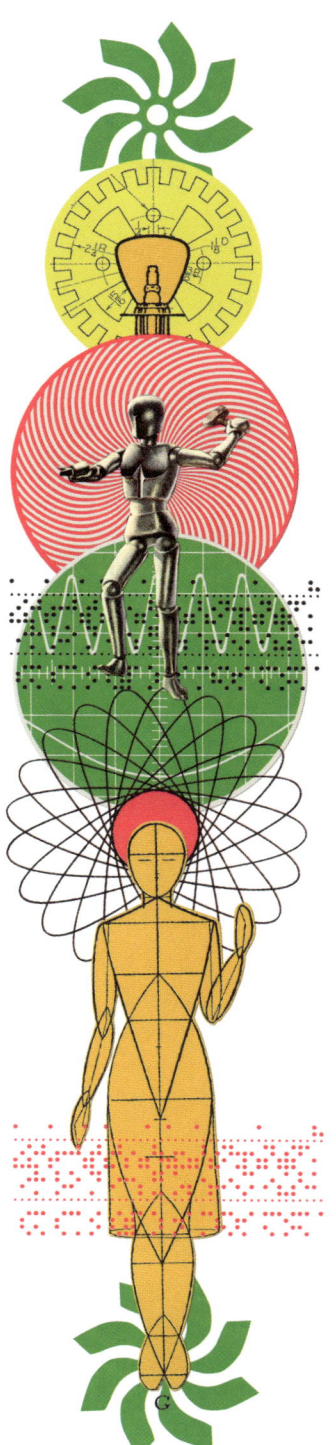

1951

RESEARCHER:
Marvin Minsky

SUBJECT AREA:
Neural computing

CONCLUSION:
A brainlike computer can "learn" like a living creature

WHAT IS SNARC?

THE FIRST NEURAL NETWORK MACHINE THAT "LEARNED" LIKE A HUMAN BRAIN

When Stanley Kubrick wanted to design the rogue artificial intelligence HAL for his 1968 film *2001*, he dreamed of "recreating," as accurately as possible, what an artificial intelligence might be able to do thirty-three years in the future (the film is set in 1991). The expert he spoke to was Marvin Minsky, who offered advice on what the machine might be able to do (in the film, it can talk and even lip read, as well as play chess), and on what it might look like (a cupboard full of black boxes).

Minsky was a visionary who, while an undergraduate at Harvard in the 1940s, had first imagined a machine that could "learn." A polymath, Minsky had studied music and biology as well as mathematics, before finding his true calling with machine intelligence. "Genetics seemed to be pretty interesting, because nobody knew yet how it worked. But I wasn't sure that it was profound. The problems of physics seemed profound and solvable. It might have been nice to do physics," he told the *New Yorker* in 1981. But neither of these had for him sufficient depth the way machine intelligence did. "The problem of intelligence seemed hopelessly profound. I can't remember considering anything else worth doing."

Inside Our Cells

Minsky had become fascinated by a paper from 1943, written by neurophysiologist Warren McCulloch and mathematician Walter Pitts, which explored how neurons (brain cells) might work. The paper modeled the idea using simple electrical circuits.

In 1951, Harvard psychologist George Miller offered Minsky the chance to build a similar machine, securing funding for him to create the device. Minsky recruited graduate student Dean Edmonds, but warned him that he feared the machine might be "too hard" to build.

In fact, the machine would become the first electronic learning system that simulated the function of a neural network. Widely used today, neural networks are computer networks that mimic the structure of the human brain.

Minsky's machine, known as SNARC (Stochastic Neural Analog Reinforcement Computer), had forty synapses built from tubes, motors, and clutches (plus a spare part from a B-52 bomber control panel).

Today, all that remains of SNARC is one of its neurons (itself a huge gadget with vacuum tubes, wires, and a capacitor) that was connected to forty others via a plugboard. The full array was around the size of a grand piano.

The idea of SNARC was to reinforce "positive learning." The machine had "memory" in the form of capacitors (components that can store electrical charge, and be used for short-term memory) and a potentiometer (used in volume controls and for long-term memory).

If a neuron fired, a capacitor retained the memory that it had done so. If the system was "rewarded" (by the researchers pulling a button), a chain connected to the potentiometer for all forty neurons would increase the future probability of the neuron firing. The combined effect of these was to "reward" correct decisions.

Rat in the Machine

Minsky tested the machine by playing the role of a "rat" trying to search a maze for food. It's not fully clear how he tracked the results, as the full machine is lost. After building SNARC, Minsky loaned the machine to students at Dartmouth, but when he asked for it back ten years later, it had disappeared. It's thought that Minsky and Edmonds tracked their progress using lights.

After several attempts, the machine would "think" logically, based on the reinforcement of correct choices, Minsky said. This meant that the "rat" would at first proceed randomly, but "correct" choices would mean the machine would find it easier to make the same choice again.

Then Minsky noticed something else: "It turned out that because of an electronic accident in our design, we could put two or three rats in the same maze and follow them all. The rats actually interacted with one another. If one of them found a good path, the others would tend to follow it. We sort of quit science for a while to watch the machine. We were amazed that it could have several activities going on at once in its little nervous system."

Machines with Brains

Minsky later pointed out some of the limitations of the nascent field of neural network research in his 1969 book *Perceptrons*, co-written with Seymour Papert. At the time, the book was blamed by some for diverting research funding away from the topic.

In recent years, however, artificial neural networks have become popular, and are now widely used in "deep learning," with computer networks composed of layers of nodes that are trained on examples (for instance using labeled images) and then used to recognize further examples themselves.

They are widely used in speech recognition and translation software, among other areas. When Google's DeepMind AI beat the world's best player at the ancient board game Go (see page 155), DeepMind used a "neural net" to learn how to play better than the best human players and devised entirely new strategies for the game.

Google has even experimented with using neural nets to design new chips for artificial intelligence, which, going back to Stanley Kubrick, HAL, and *2001*, sounds like a chilling tale from science fiction.

WHEN WAS ARTIFICIAL INTELLIGENCE BORN?

THE DARTMOUTH CONFERENCE

1956
RESEARCHER:
John McCarthy
SUBJECT AREA:
Artificial intelligence
CONCLUSION:
Defined the challenges of AI (and spawned the field)

The term "artificial intelligence" was coined in August 1955 in a proposal for a workshop about "making intelligent machines." The proposal, submitted by then assistant mathematics professor John McCarthy of Dartmouth College, New Hampshire, highlighted the optimism that many scientists felt in the early 1950s: that artificial intelligence was not an intractable problem, and one that might be achieved in the near future. Reading the language of those papers now, it sounds as if AI was something that might be achieved, at the latest, by the end of the decade.

McCarthy is widely believed to have been the one to coin the phrase "artificial intelligence," which he defined as "the science and engineering of making intelligent machines." His idea was that at the conference, "an attempt will be made to find how to make machines use language, form abstractions and concepts, solve kinds of problems now reserved for humans, and improve themselves.[...] For the present purpose the artificial intelligence problem is taken to be that of making a machine behave in ways that would be called intelligent if a human were so behaving."

Thinking Machines
Attended by around fifty academics, including Marvin Minsky, inventor of the first neural net device (see page 86), the workshop ran through July and August of the following summer. It's generally considered to be the birthplace of artificial intelligence as a field, and many of those in attendance—mathematicians and scientists— went on to have their own breakthroughs in AI.

But the language of the proposal highlighted the fact that many AI luminaries had an unrealistic optimism, believing that a computer could achieve humanlike feats of intelligence within the near future. Even now, more than six decades later, the predictions

of the proposal have not come to pass.

While AI and machine-learning systems can do humanlike things such as talking in natural language, the systems are not intelligent in the way that many of the attendees at the Dartmouth Conference imagined.

The proposal said: "We think that a significant advance can be made in one or more of these problems if a carefully selected group of scientists work on it together for a summer."

The "problems" that the researchers blithely hoped would be solved included computers simulating human brains, neural nets, computers using language, and self-improving machines.

One proposal read: "The speeds and memory capacities of present computers may be insufficient to simulate many of the higher functions of the human brain, but the major obstacle is not lack of machine capacity, but our inability to write programs taking full advantage of what we have."

The AI Winter

The idea that it was possible to create an artificial intelligence simply by writing clever software for expensive, slow 1950s computers was spectacularly wrong. So were many of the other predictions of the Dartmouth Conference.

In the 1960s and 1970s, the growing power of computers (and falling prices) meant that interest in artificial intelligence remained high. But the failure of the field to deliver anything resembling a true artificial intelligence (or indeed a machine that could understand language, or improve itself) meant that funding for the field dwindled in the later 1970s and 1980s, leading to what has been described as an "AI winter."

In 1973, Professor Sir James Lighthill was tasked by the U.K. Parliament to evaluate the state of artificial intelligence research in the United Kingdom. His report criticized the failure of AI to deliver on its "grandiose objectives." "In no part of the field have the discoveries made so far produced the major impact that was then promised," he wrote. His report suggested that AI algorithms were not capable of dealing with real-world problems, and it led to cuts in funding for research first in Britain, then the United States.

Interest in "artificial intelligence" would rekindle in following decades, but without the blind optimism that led the Dartmouth attendees to believe that creating humanlike intelligence was a problem that could be solved by a small number of scientists over a hot New England summer.

Machine Philosophy

McCarthy would go on to contribute to the philosophy of artificial intelligence, writing, "Machines as simple as thermostats can be said to have beliefs, and having beliefs seems to be a characteristic of most machines capable of problem solving performance."

He was disappointed by systems such as Deep Blue, the supercomputer that beat Garry Kasparov at chess (see page 117), believing that AI research had focused too much on simply dealing with the same problems, but faster and faster.

Colleague Daphne Koller said that in his later years McCarthy (who died in 2011) still hoped that there would one day be a machine that could pass the Turing Test, rather than the narrow, super-charged approach of modern AI. "He believed in artificial intelligence in terms of building an artifact that could actually replicate human-level intelligence."

1960

RESEARCHER:
John Chubbuck

SUBJECT AREA:
Learning robots

CONCLUSION:
A robot can "feed" itself independently

CAN A MACHINE LOOK AFTER ITSELF?

HOW THE BEAST LEARNED TO FEED ITSELF

Although it would be three decades before NASA's Mars Sojourner robot would explore another world, at the dawn of the 1960s, experts at John Hopkins University in Baltimore, Maryland, were already pondering how to create robots that could survive on their own.

It wasn't quite the surface of Mars, but in the markedly less hostile environment of the corridors of John Hopkins's Applied Physics Laboratory, two robots, The Beast and Ferdinand, were designed to survive by themselves.

A "Weird-looking Monster"

The definition of "survival" was: not to get lost, not get stuck on any objects, and to ensure that they remained charged up. They could achieve this by themselves, using sensors to hunt down electrical power sockets. The record, held by a modified version of The Beast, was forty hours without human input, one of the researchers said, with the robot's run only ended by a mechanical failure. The two-foot-high (60 cm) robot explored its world with one "arm" extended to feel its way along a wall, as if it was a lost human trying to navigate a maze. Experts in the laboratory hoped that the machines would be the foundation of robots that could explore the depths of the ocean and other planets in the solar system.

John Hopkins robotics expert John Chubbuck, who would go on to play a role in designing the guidance systems that sent Apollo missions to the Moon, described Ferdinand as a "weird-looking monster" and considered the transistors and microswitches that controlled it as a "simulated nervous system." In demonstrations, he showed how The Beast and Ferdinand could survive in messy office environments ("This is a very cluttered environment," he laughed), moving across doorways and through offices filled with chairs.

Each machine was equipped with sensors to allow it to navigate and find electrical outlets on the wall from which to charge itself. Once charged, the machine would switch into a different mode and set off to explore again.

The robot had twenty-one different modes of operation including sleep, feed, high speed, and low speed, and could be controlled from a console. The Beast Mod II weighed in at 100 pounds (45 kg), and was just under 20 inches (50 cm) wide. Inside it were 150 digital circuits and servo motors, which allowed it to extend its prongs and charge itself.

Navigating Like Bats

If The Beast became trapped against a wall, it would switch into "vibratory" mode to help it escape. To help it "sense" the world around it, it had an array of microswitches that helped it guide its charger prong into the right place. If it failed on a first attempt, it would try again, then switch back to navigation mode to find another outlet. Like a bat, it could navigate using acoustics, to enable it to move down corridors without touching the walls, with two side beams where it would measure return times for the sound, so that it could stay in the middle of the path.

An optical system allowed it to recognize the black cover plates of the wall plugs dotted around the laboratory, although Chubbuck admitted that it was prone to mistaking anything roughly the same shape for the charging panels, including chair legs.

While both Ferdinand and The Beast could be "driven" from a desk, the machines were fully independent. But, unlike later robots, they could not learn from their environments.

Not that there was no learning going on. In a video showing off the abilities of the robot, the Applied Physics Laboratory wrote, "Although the automaton is not learning from its environment, its designers are learning from the automaton." The researchers hoped to add further sensors to the robot in an effort to build something that could be used as an explorer robot in hostile environments.

The Beast is often described as a "pre-robot": it's a cybernetic system, similar to a classic thermostat-heater combination. Just as a thermostat sets a "goal" of reaching a particular temperature, The Beast's onboard electronics set it the "goal" of finding charging points and recharging its batteries. It had no computer on board and no programming language.

Lack of Interest
John Hopkins researcher Ronald McConnell wrote to *Scientific American* to say that while the robot drew some press interest, including a brief feature on NBC, government agencies, including NASA, didn't trouble themselves with it. "ARPA (Advanced Research Projects Agency) came by, but in the era of early near-Earth, manned space flight, it wasn't really interested in prototypes of robots to explore the Moon, Mars, or the undersea on Earth," he wrote. "Johnson Wax wanted to know if a robot floor waxer was feasible."

Today, similar systems allowing robots to navigate back to their charging pad under their own steam have become a familiar part of devices such as robot vacuum cleaners. Even Honda's humanoid ASIMO robot gained the ability to find his own charger, as have other "toy" robots such as Anki Vector.

CAN A ROBOT DO A HUMAN JOB?

HOW ROBOTS REVOLUTIONIZED MANUFACTURING

1961
RESEARCHER:
George Devol
SUBJECT AREA:
Robot arms
CONCLUSION:
Robots revolutionized manufacturing

At a cocktail party in 1956, two American engineers chatted about their shared interest in science fiction novels—and, in particular, the *Robot* novels of author Isaac Asimov, with their robot servants, and Asimov's "three laws of robotics," designed to stop robots harming their human masters. In his books, including *I, Robot*, Asimov portrayed a distant future in which benevolent robots worked alongside human beings.

One of the men, George Devol, explained that he had patented an idea for a Programmed Article Transfer device. The other engineer, Joseph Engelberger, exclaimed, "Sounds like a robot to me!"

Engelberger licensed Devol's patent, which ultimately became the Unimate, the first robot arm used on a production line. It is similar to models still used today.

Between them, the two men would reshape the world of manufacturing forever, but having partnered to create the robot, they at first faced incredulity and hostility from the companies they pitched it to. Many simply didn't believe that such a device was possible. Engelberger approached forty companies before he persuaded someone to invest in the machine. "To try to get a normal businessman to understand a robot . . ." Devol said. "They thought you were talking about science fiction or something like that." The patent for the robot wasn't granted until 1961, five years after the two had met. They finally sold the

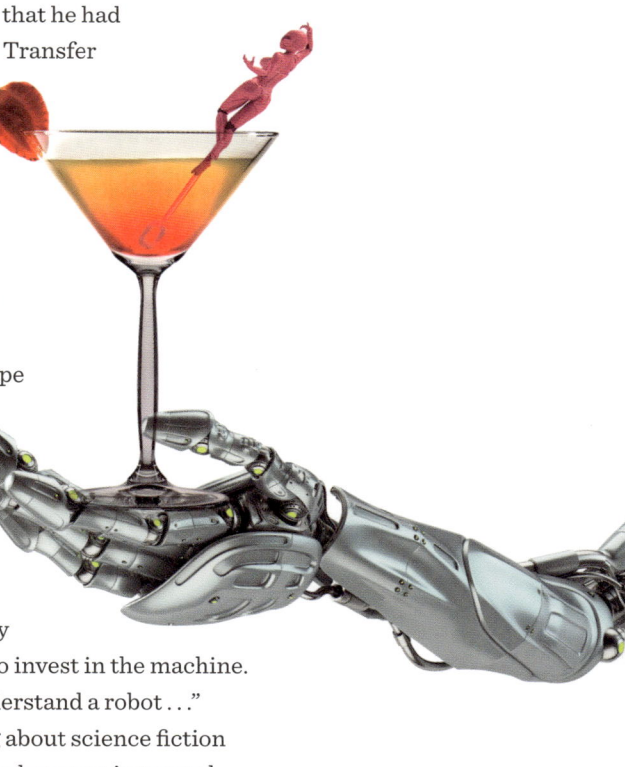

first Unimate robot to General Motors.

The first Unimate's "job" was lifting and stacking hot metal parts at the General Motors plant in Ewing Township, New Jersey. It was a dangerous and unpleasant task for human beings, but one that the programmable robot arm could do effortlessly.

Replacing Jobs?

Quickly the Unimate 1900 series went into mass production, and there were soon more than 400 of the robotic arms at work in the United States. The robot mesmerized the world. Appearing on TV on *The Johnny Carson Show,* it potted a golf ball and poured a beer. It also attempted to play the accordion, which was much less successful. Carson marveled that the machine "could replace someone's job."

The Unimate was programmable, with a magnetic drum that could store instructions. There were no sensors on the device. All it could do was repeat the same task over and over again.

Chrysler and other companies bought more Unimates (with new models handling tasks including welding and spray-painting), and the technology became popular in Japan, helping the Japanese car industry to leap forward on the global stage.

In the following decades, Japan and then China became enthusiastic users of robotics. According to the International Federation of Robotics, there are 2.7 million industrial robots operating in factories around the world today, while American science title *Popular Mechanics* named the Unimate robot arm one of the top fifty inventions of the twentieth century.

Hot Dogs and Burgers

In the 1940s, the self-taught Devol had invented a version of a microwave oven, a coin-operated machine he dubbed the Speedy Weeny, which doled out cooked hot dogs. In his house, his wife cooked burgers with a similar machine that Devol had created. He had also invented an automatically opening door, which was advertised as the Phantom Doorman. In all, Devol amassed more than forty patents during his life.

In a later interview with *Computer World*, he said that his self-taught background had never held him back. "I always went into areas of industry where nobody else knew anything either," he said. "There was nowhere to go to get information, so I generated it."

Engelberger and Asimov

Engelberger would go on to become known as "the father of robotics," not just a pioneer of the technology but a tireless advocate for the use of robotics in everything from hospitals to space exploration. He offered advice to NASA on using automation in space missions, and would work on robots built for hospitals, with his HelpMate hospital courier robot widely used.

He later thanked Asimov for beginning his prolific writing career at just the right time to inspire Engelberger when he was a physics undergraduate at Columbia University. Engelberger's own book, *Robotics in Practice*, came, appropriately, with a foreword by Asimov, in which the novelist wrote: "Does a robot displace a human being? Certainly, but he does so at a job that, simply because a robot can do it, is beneath the dignity of a human being; a job that is no more than mindless drudgery. Better and more human jobs can be found for human beings—and should."

CHAPTER 5: Survival of the Fittest
1970–1998

Can robots learn new tricks from living things? Some researchers during the 1980s began to believe that robots could behave more like animals, such as insects—or even human beings. Robots such as Toto learned to explore their environment using a simple, ratlike "brain," while researcher Cynthia Breazeal developed the first "social robot," built to respond to emotions like a small child (and with its own, childlike needs).

In a huge tank at MIT, a robotic tuna named Charlie swam endlessly against a current, allowing researchers to learn how

real fish propel themselves through the water (and design new machines for exploring beneath the sea).

    Meanwhile, other robots would face human challenges, with Honda's iconic ASIMO becoming the first robot to walk like a human, and teams of robot soccer players embarking on a mission to beat the best human team in the world by 2050. By 1997, IBM's cupboard-sized Deep Blue computer was about to win a chess game that marked a turning point in the history of AI . . . and the human race.

1970

RESEARCHER:
Charles Rosen

SUBJECT AREA:
Robots that navigate

CONCLUSION:
A robot can navigate and deal with obstacles by itself

HOW DID SHAKEY THINK?

WHY SHAKEY'S NAVIGATION CHANGED THE WORLD

Today, most of us rely on computers to tell us where to go, and do so without a second thought, thanks to applications such as Google Maps built into every smartphone.

But the idea of a computer that could navigate by itself was cutting-edge in 1964, when Charles Rosen, head of the Machine Learning Group at the Stanford Research Institute in Menlo Park, California, suggested it to ARPA, the research arm of the U.S. Department of Defense.

Robots that could find their way around by themselves had previously only existed in science fiction. Angling for funding, Rosen suggested that the robot could "perform reconnaissance missions" that would normally require human intelligence. ARPA was interested, and in 1966 provided support for the project.

The researchers suspected that the military hoped the technology might lead to a robot that could count Chinese tanks. That never happened, but, in many ways, Shakey—so named because the tower of components and video cameras wobbled when the robot moved—was the first machine similar to what most people understand today as a "robot."

Sparking debate around robots and AI, it became an iconic figure in media, in much the same way as later robots such as ASIMO (page 127) became celebrities in their own right.

"In an antiseptic windowless laboratory in this balmy California town, an ungainly automaton is taking its first baby steps

in learning to perform complicated tasks on its own," wrote *The New York Times*. "It's still a 'very dumb machine' according to its engineer 'parents.' All it can do is move from one point to another through a room full of obstacles with only a feeble 'awareness' of its environment." Where *The New York Times* likened Shakey to a "baby" learning by itself, *Life Magazine* described it as "the first electronic person."

In a promotional film, the robot's team said: "Our goal is to give Shakey some of the abilities associated with intelligence, abilities like planning and learning. The main purpose of our research is to learn how to design these programs so that robots can be employed in a variety of tasks, ranging from space exploration to industrial automation."

A Red or White World
Shakey could "see" using video cameras, "feel" using cat whisker sensors, and navigate through a laboratory by itself, with a maze of isolated prismatic blocks that resembled a children's soft play area. Everything in Shakey's world was painted white or red, in order to make things clearer for the robot's monochrome vision,

while still reflecting enough light for its laser range finder to work.

It communicated with researchers via radio and moved using a set of wheels controlled with motors. It was also armed with a push bar that could move blocks in front of it. Peter Hart, one of the researchers behind Shakey, described it simply as "an electronics rack on wheels."

But Shakey was the first robot that could perceive and plan. The key to its unique abilities was that the "thinking" was not done inside the washing-machine-sized unit itself: Shakey was linked to a PDP-10 computer weighing several tons, which processed the data from its sensors and sent commands to the motors that moved the wheels.

Dead Reckoning

The robot navigated by "dead reckoning," that is, by counting the rotations of its wheels, but could back this up by using his camera to "see" where it was, building up a simple map of the laboratory it was in. It could respond to simple commands such as, "ROLL" and "TILT" and could also be commanded to "GOTO" a specific location in the lab. The robot was given commands by teletype (an electromechanical keyboard) and displayed what it was doing via a cathode ray tube (an old-style television).

But what marked Shakey out was its ability to cope with unexpected obstacles. In the film, "Charlie the Gremlin"—in other words, Charles Rosen wearing a cape—Rosen deposits a box in the robot's way to represent an unexpected event. Shakey "sees" the box and assesses what it is, then modifies its plan to divert away, trundling round to its goal from a different direction. Researchers could "see" the robot's thinking on a screen.

The robot's STRIPS planning software (Stanford Research Institute Problem Solver) allowed it to deal with "missions" that involved pushing blocks and flicking light switches. "If Charlie the Gremlin came and did something upsetting," Nils Nilsson, who worked on the project, said, "STRIPS could cook up a new plan. It was a really complex program for its day."

Shakey was capable of locating a specific spot in his environment, which was made up of seven linked rooms. It could also find designated boxes and push them with its push bar into groups, under instruction from human researchers (while avoiding obstacles in its path, whether they were part of the environment, or placed there by Charlie the Gremlin).

Spin Me Right Round

The robot did, however, have its eccentricities. "Shakey would sometimes stop what it was doing and start pivoting around 360 degrees around and around," said Peter Hart. At first, this left the scientists perplexed. But, Hart explained, "we dug down in the code and found that there was a routine in there intended to unwind cable." Initially, the robot had been attached to a long wire: hence it was programmed to unwind itself.

The project was eventually canceled by ARPA—Nils Nilsson said the defense agency's words were "no more robots"—but Shakey's methods of navigation and planning would have an impact on robots for the next fifty years, influencing everything from video games to Mars rovers.

Computing methods devised to help the robot navigate its brightly painted block mazes are still in use in self-driving car software today, and when you ask your phone for driving directions, it will use algorithms designed for Shakey.

"The holy grail of software is Artificial Intelligence, either in a pure software capability or in a physical robotic capability," Bill Gates said. "Going back to the Sixties, the Stanford Research Institute (SRI) had their robot, Shakey. I remember seeing that and saying, 'That's what I want to work on—making that robot a lot better.'"

Now "retired," Shakey is on display in a glass case in the Computer History Museum in Mountain View, California.

1987

RESEARCHER:
John Adler

SUBJECT AREA:
Radiosurgery

CONCLUSION:
Robotic cancer treatments have saved thousands of lives

CAN ROBOTICS BE USED TO TREAT CANCER?

CYBERKNIFE RADIOSURGERY

Dr. John Adler said that he treated the development of the CyberKnife robotic radiosurgery system as if he were performing brain surgery—and everything was going wrong. He had to force himself to stay positive and just put one foot in front of the other. Creating CyberKnife proved, however, to be a far longer journey than any brain operation.

Adler, an American neurosurgeon, was aware that his colleagues at Stanford University thought his proposed design for a robotic radiosurgery device would come to nothing, and had described it as "Adler's folly."

Then when he pitched the idea to venture capitalists, they were shocked by the device's size—it was 7 feet (2 m) tall—and the fact that each unit would cost $3.5 million. "No one really believed that it was economically viable or medically superior," Adler said. "It just fell on deaf ears."

Robot Surgeon

But the CyberKnife would go on to save thousands of lives, and would fundamentally change the way some cancers were treated. The device, which is now installed in dozens of hospitals and medical centers worldwide, is a robotic radiosurgery system that captures images of patients' bodies as they are treated, meaning that the CyberKnife can operate incredibly precisely, delivering radiation from multiple angles even, to tumors that would normally be untreatable.

The machine's linac, or linear accelerator, is directly mounted on a robot arm, delivering the high-energy X-rays or photons used in radiation therapy. It can even synchronize with the patient's breathing to ensure that radiation is delivered in the right place.

But back in 1987, when Adler first came up with the idea, the technology required to deliver on the promise of the CyberKnife barely existed, and developing the device proved to be an engineering nightmare.

On a fellowship in Sweden in 1985, Adler had been inspired by Professor Lars Leksell, the inventor of radiosurgery, who had created a device called the Gamma Knife. Looking faintly like a medieval instrument of torture, this was a metal frame that surrounded a patient's head to guide radiation beams.

Leksell himself had faced opposition to his ideas, but believed that there had to be an alternative to traditional surgery. "Tools used by the surgeon must be adapted to the task and where the human brain is concerned, they cannot be too refined," he said.

The Gamma Knife was cumbersome to operate and time-consuming to set up. But it worked. Adler said that after seeing patients walk out of hospital two days after treatment, with no scars, he realized that this was the future. His idea, which would take almost two decades to make a commercial reality, was to harness the emerging science of robotics to further refine the Gamma Knife.

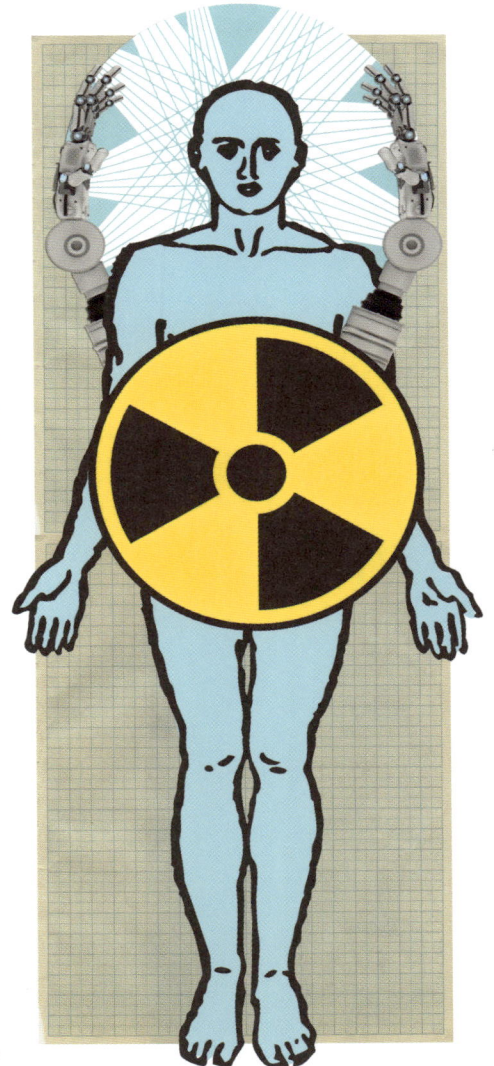

Taking it Further

Adler's own CyberKnife, the details of which he worked out with engineers at Stanford on his return to the U.S., was guided by software, with a nimble robot arm moving around the patient to deliver precisely targeted blasts of radiation. Or at least that was the theory.

Early tests of the system were not immediately successful. An elderly woman with a brain tumor was treated with the frameless device, but software bugs meant that the surgery lasted almost a whole afternoon. "By most measures, frame-based radiosurgery would have been vastly simpler," Adler admitted. "But we had made the first clinical step."

The woman sadly died shortly afterward, not surviving long enough to have a follow-up MRI scan. The cause of her death was unclear.

The technical problems that Adler faced were immense, and he battled alongside engineers to iron out bugs, treating just one patient a month in the early days of the CyberKnife, when there was only a single unit installed in Stanford itself.

Growing Pains

Then, when he launched Accuray, a company to market CyberKnife, that too faced a series of disasters. At Christmas 1994, a potential buyer pulled out and early the following year the company ran out of money, leading to two-thirds of the employees being laid off.

In 1999, Adler took over as CEO. "People were fighting. It was ugly," he said. "We had no money and everybody hated each other and our customers hated us. There was really nothing going for us." But around that time, America's Food and Drug Administration (FDA) cleared the CyberKnife first for use on brain tumors, then for tumors anywhere in the body.

Accuray slowly found and retained customers, and continued to develop new systems, which were sold to hospitals around the world. Today Adler is credited with creating a whole field of image-guided radiation targeting (IGRT).

The latest CyberKnife S7 can synchronize in real-time with patient movement, and can deliver radiation from thousands of unique angles, with sub-millimeter accuracy and no input from a human surgeon.

The CyberKnife has now been used to treat more than 100,000 patients worldwide. In general, robots are increasingly used in surgery, particularly on "minimally invasive" procedures and keyhole surgery. Meanwhile, other technology companies are working with robots that can perform surgery remotely, meaning that surgeons can operate on patients even when they are on a different continent.

CAN MACHINES LEARN FROM THEIR BEHAVIOR?

HOW TOTO HELPED MACHINES TO "LEARN"

1990

RESEARCHER:
Maja Matarić

SUBJECT AREA:
Behavior-based robotics

CONCLUSION:
Robots can learn to navigate using a brain like a rat's

Could a robot's control system help it to form a map of its surroundings—in the same way as a rat forms a map in its brain? This was something that had never been achieved in robotics, but Toto, a robot built at MIT in the early 1990s, could not only "map" areas by itself, but revisit previous landmarks, with an approach to navigation not dissimilar to a lab rat in a maze.

Built by roboticist Maja Matarić, Toto had a "layered" control system, which allowed it to simultaneously have a "primitive" layer of commands, enabling it to wander randomly through its environment, avoiding obstacles, and more sophisticated commands built on top. As the robot wandered, it used sonar and compasses to build a map of its environment, and could then work from that to reach places that it had already visited (it could accept commands via buttons on its exterior). Matarić described Toto as, "Sort of like a rat in its brain, navigating around a maze."

Bottom-up Robotics

Toto was an example of "behavior-based robotics" championed by Rodney Brooks at MIT (who would go on to work on the Roomba robotic vacuum cleaner, see page 136). Brooks promoted the idea of behavior-based systems where a series of simple "behaviors," for instance, following boundaries or staying away from messy areas, guided a robot's movements. Also described as "bottom-up" robotics, Brooks said that he was inspired in his approach by insects that can make decisions quickly despite not being

particularly intelligent. Behavior-based robots act first, and think later, allowing them to explore and achieve goals, despite not having much in the way of pre-programmed behavior (or intelligence). This is what allowed Toto to find its way.

Like other behavior-based robots, Toto had behaviors layered into a hierarchy, with higher layers overruling lower layers (for instance to guide Toto to a landmark that it had visited before).

Rat in a Maze

Robots built using this system were simple, but were capable of relatively intelligent behavior, often resembling an insect (or with Toto, a rat) in their approach to problems.

In Toto's case, it allowed the robot effectively to map the lab environments it explored. The "map" it recognized was simply what Toto had done in certain areas beforehand.

If it was walking in a straight line without obstruction for a long period, it would map this as a corridor, or if it detected a wall, it would map this as "right hand wall" or "left hand wall," and if it was wandering in a messy area, it would map this as "messy area."

When Toto's landmark-detecting layer did detect a landmark, the description was sent to all of Toto's map behaviors. If one matched, then that behavior would become active, meaning that Toto knew where it was on a map. The system also simultaneously sent inhibiting behavior to other areas, so that only one area could be active at one time, meaning that Toto would be even more certain of its location.

If no landmark matched, the control system would "create" a new area, allowing Toto to explore its mazelike world. Toto would try to predict what map area and behavior would come up next, based on its map: if it was right, the robot would be even more confident that it was in the right place.

For humans, knowing where we are is very easy, especially in environments that we navigate daily, such as our own homes or offices. For robots, it's a very tricky challenge indeed.

Navigating the World

Toto's ability to know where it was also meant it could navigate to landmarks that it had previously visited. To do so, the researchers would define a goal landmark, which would send messages to nearby behaviors on the map until they reached the behavior where Toto actually was (the robot's real location). Toto would then sort through the lists of commands until it found the shortest list, which would be the shortest route to the landmark it needed to reach.

"Besides going to a particular landmark, such as a specific corridor, Toto could also find the nearest landmark with a particular property," Matarić wrote in her 2007 book *The Robotics Primer*. "For example, suppose that Toto needed to find the nearest right wall. To make this happen, all right wall landmarks in the map would start sending messages. Toto followed the shortest path, and reached the nearest right wall in the map."

The simplicity of Toto's navigation meant that it could still find the shortest route even if it was picked up and put in a different area—something with the potential to completely baffle more complex robots. The way that the robot "learned" and internalized its map while exploring is similar to the way rats learn about their environment, Matarić believed. Such behavior-based robotics allowed for robots to achieve complex goals, such as navigating to an objective, without complex programming.

Matarić would go on to pioneer new ideas in social robotics for elderly and sick patients, while the idea of behavior-based robots is still influential, and has, since the 1980s, helped many cheaper robots, such as robot vacuum cleaners, be useful.

1990s

RESEARCHER:
Cynthia Breazeal

SUBJECT AREA:
Social robotics

CONCLUSION:
Robots can make emotional bonds with people

CAN ROBOTS EXPRESS EMOTION?

KISMET AND SOCIAL INTELLIGENCE

"No, no, not appropriate," the woman says sharply to the disembodied robot head. The robot's head slumps in evident shame, and even its ears fold down, as if it is genuinely feeling remorse. It looks like an animated character, perhaps from a Pixar film. But there is no special effects trickery at work: the head is a robot.

The robot head (Kismet) was designed in an MIT lab by roboticist Cynthia Breazeal, who said that she was inspired to investigate "social robotics" while working on NASA's Sojourner rover. Rather than focusing on how robots would get from point A to point B, Breazeal wanted to work on robots that would make people feel comfortable interacting with a machine.

Brought up by two scientists, Breazeal believed that social robotics was something most roboticists simply didn't think about. Her own interest in "social robots" began after having written a short story about a robot with emotions as a young child. Her idea for what such robots would be like was inspired by fictional robots including R2-D2 and C-3PO from the *Star Wars* films.

"There are people and pets, minds, thoughts and beliefs, and emotions," she said, "and robots need to be able to interact with them. What would it mean to build a robot with social and emotional intelligence that can ultimately do things in collaboration with people?"

Friendly Robots

Today, most of us talk to social robots such as Siri and Alexa without really thinking about it. In everything from banking to ordering a pizza, such "bots," which mimic the way real humans speak and act, are becoming ubiquitous. Most of us even expect robots and AI agents such as Siri to emote and use colloquial language when we address them.

But before Kismet, Breazeal pointed out, roboticists had not seriously addressed the fact that robots would need to deal with thoughts, beliefs, and emotions, or that they would need some form of social intelligence.

The approach Breazeal and her team took with Kismet was unique. Instead of being pre-programmed, it would learn like a human baby does, by paying attention to a parent.

Kismet can't actually understand language, but it can interpret the intent of someone who is speaking. Nor does it actually speak in any intelligible language, instead making wordlike burbles. The robot would learn, Breazeal hoped, through the same sort of exaggerated gestures parents use to a small child, and would respond in kind, reacting to the intention of the person talking to it.

As If Alive

The result is a robot that appears to have some measure of social intelligence, and to respond like a living creature. It perceives the world through video cameras and microphones, and can react using motors that move its head, ears, and lips.

Kismet looks like a toy (or a film prop) and has inspired generations of toys such as Furbies, but on the inside it's a huge amount of cutting-edge computer hardware. One system handles speech synthesis and intent recognition (Kismet's ability to "understand" the emotional intent of people talking to it), running on two Windows PCs and one Linux machine. Separately, four Motorola microprocessors handle perception, motivation, motor skills, and facial movements, while another system with nine networked PCs handles vision

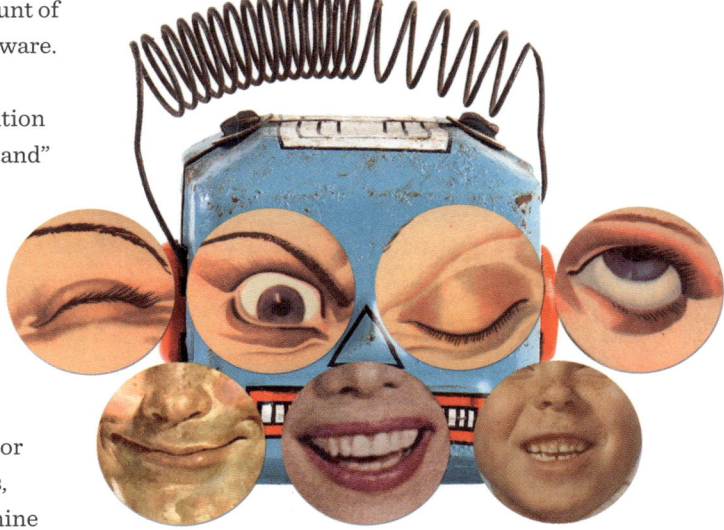

processing and eye and neck control.

In simple terms, the robot processes images and sounds, looking for things to react to (such as tone of voice, or whether someone is looking at it), and then feeds this to an attention system that directs Kismet to pay attention to something.

If it detects a human, it has a wide variety of emotions, from happiness to disgust, plus reactions such as boredom. Many of its reactions are designed to "control" the human that it's interacting with: for instance, if someone is too far away for Kismet's cameras, it will make a "calling" noise to entice them nearer.

Robot Desires

But the robot also has its own needs. A computer attached to Kismet shows its three "drives" (social, stimulation, and fatigue) on a bar chart, with each drive a "need" that it attempts to satisfy. If it's lonely (i.e., if its social drive is high), it seeks out human interaction. If it's bored or in need of stimulation, it will stare at a toy, hoping for someone to bring it over. When tired, it wants to rest.

The result of all this computing power was that Kismet—a disembodied head—could "intuit" emotions and react with its own. When "surprised," the robot would lift its ears and open its lips. When "disgusted," it would tighten its mouth. When sad, its ears would slump downward, and its mouth went into a cartoonish grimace.

Breazeal later developed other "social" robots including a robotic diet and exercise coach, and "telepresence" robots that allow people to "send hugs" over long distances. She also founded the social robotics company Jibo.

The world is, she believes, on the verge of "social robots" becoming a common presence in every home. "As mobile computing develops, and the cost of sensors, processors, and wireless communication falls, domestic-service robots will become a reality. Social robots won't replace human networks, but will supplement and strengthen them."

CAN ROBOTS SWIM UNDER WATER?

HOW ROBOTUNA HELPS US EXPLORE THE SEAS

1993
RESEARCHER:
Michael Triantafyllou
SUBJECT AREA:
Robot propulsion
CONCLUSION:
By imitating animals, robots can swim quickly and efficiently

When humans design a system for underwater propulsion, they're competing against fish, which have been "designed" by evolution over a span of 160 million years. So why, wondered MIT Professor Michael Triantafyllou, had no one attempted to learn from how fish swim through the water?

When MIT built RoboTuna, it was entirely unique. No one had tried to replicate the movement of fish before. The team chose tuna for their first robotic fish because of their speed. Tuna have evolved to cut through the waves at enormous speeds, with a specialized body shape that enables some tuna species to hit 43 miles per hour (69 kph). Bluefin tuna can reach up to 10 feet (3 m) in length, and weigh more than a horse.

At MIT, they described their work as akin to "reverse engineering," as the team sought to imitate the speed and movement of the bluefin. The robot (which swam attached to a strut in a huge tank, with wires feeding back information to his creators) became affectionately known as Charlie.

Fishy Tale

Charlie had a skeleton of aluminum lined with forty polystyrene ribs, and was wrapped in a skin of reticulated foam and Lycra, which helped to smooth its passage through the water. Unlike every previous human-designed water vehicle, its method of propulsion wasn't paddles, or a sail, or a propeller ... it was a fin.

The robot had around 3,000 components, and its body expanded and contracted to order using six servomotors, which were each rated at two horsepower. The motors were attached to a system of stainless steel cables within Charlie's body, like muscles and tendons.

On the outside of Charlie, force sensors mounted on the robot's ribs provided feedback, allowing it to adjust its motions in real time. The machine took several weekly swims in MIT's tow tank, with the researchers measuring the feedback from Charlie to understand, for the first time, how tuna swam. The data from Charlie would allow the researchers to imagine a new way of propelling undersea vehicles.

Vortex Master

The researchers found that controlling eddies (or vortices) in the water was crucial to how fish swam (and very different from how human-made vehicles had, until that point, propelled themselves). Tuna propel themselves by manipulating eddies in the water, and creating their own by moving their tails.

Professor Triantafyllou wrote at the time, "Current technology aims to minimize the vortices formed by moving through water, since these cause an enormous amount of drag, which slows the vehicles down. Fish, however, purposely create these vortices and exploit them for their own benefit. This is what we're doing with the RoboTuna. We want to create the vortices but we want to control them."

The researchers used a "genetic algorithm" to "evolve" Charlie's swimming system, selecting programs that performed better and better. Over time, Charlie was able to master vortices and replicate (to some extent) the bursts of speed that real-life tuna are capable of (albeit while still tethered to a pole in its tank at MIT).

Exploring the Deep

The researchers hoped that Charlie's technology could be used for future responsive underwater vehicles, designed to work in extreme environments. "When exploring thermal vents at the sea floor, the water temperature can vary by 210 degrees Fahrenheit (100°C) within just a few feet," Triantafyllou said. "Because of this, you need a system that is flexible, and that can react extremely quickly to unforeseen occurrences. Current autonomous underwater vehicles (AUVs) do not have the kind of speed and agility that such dangerous situations require, and so many are lost to unforeseeable circumstances. […] RoboTuna will minimize risk in exploring areas that are being covered now by the clumsy traditional propeller-driven AUVs, as well as open up new areas that have until now been considered too dangerous."

RoboTuna's breakthrough inspired a vast number of robotic fish to be built in labs around the world— with MIT building a RoboPike, in an effort to understand the furious acceleration of real-world pike, and to investigate Gray's Paradox, posed by British zoologist James Gray in 1936, about how dolphins can swim at the speeds they do while seeming not to have enough muscles. There  have been several dozen other robotic fish built in the wake of Charlie the RoboTuna.

In 2009, MIT researchers created a new generation of robotic fish, much smaller than RoboTuna, at 5 to 18 inches (13-46 cm) in length, each built from a soft polymer that could resist corrosion, even after long periods fully immersed in water.

The fish have just ten parts, compared to the thousands inside RoboTuna, and cost a few hundred dollars each, with companies interested in using the devices for underwater measurements and surveillance. The idea is that hundreds of the relatively cheap devices could be thrown into a bay or a port, then take measurements.

With fish easily spooked by human presence, robotic fish could also allow humans to observe the animals without being spotted. An MIT team made a soft-bodied robotic fish that swam alongside real fish in Fiji's coral reefs . . . without being rumbled as a robot impostor.

1997

RESEARCHERS:
Hiroaki Kitano et al.

SUBJECT AREA:
Robot challenge

CONCLUSION:
A robot team could defeat the best human team by 2050

WHO IS BETTER AT SOCCER?

THE GOAL OF THE ROBOCUP

By the year 2050, a team of robot soccer players will have beaten the best human team on planet Earth, adding soccer to the list of human endeavors where machines have triumphed over the human race forever (like chess, see page 119). Or at least, that's the theory.

The official goal of RoboCup is: "By the middle of the twenty-first century, a team of fully autonomous humanoid robot soccer players shall win a soccer game, complying with the official rules of FIFA, against the winner of the most recent World Cup."

From the early 1990s, robot experts had suggested that a useful "grand challenge" event might be to attempt to create a robotic soccer team capable of taking on humans. Considering the difficulty of even getting robots to navigate a field, let alone work as a team to defeat highly skilled human players, it was a very steep challenge indeed. Initially supposed to be limited to Japan, the idea attracted so much attention from around the world that it was thrown open, and the RoboCup was born.

Aiming for Goal

When the RoboCup first began, it was difficult to envision the robot "players" on the field as being able to mount a serious challenge to the best human teams. Even making contact with the ball was a stretch, never mind outmaneuvering defenders or aiming at a goal.

Launched in 1997 by experts including Hiroaki Kitano, a research scientist at Sony who worked on its iconic robot dog Aibo, the RoboCup saw teams of robot and artificial intelligence researchers descend on Nagoya in Japan to pit their wits against each other in several different leagues (divided by the size and

ability of the robots). One of the inspirations for the competitors was Deep Blue's chess victory over Garry Kasparov in May 1997 (see page 119).

The rules are simple. The robots have to be fully autonomous: no human control from the touchline, and in fact no human interaction at all once the starting whistle blows. Kitano recounts that at the first RoboCup, two teams of robots were on the turf, looking over their field with sensors and only moving slightly. A reporter asked him when the game was about to begin. "It started five minutes ago!" Kitano said.

The machines had taken several minutes just to orient themselves and work out what to do next. Another early game saw one team "win" because they were the only one to have made contact with the ball.

Dogs on the Field

Changes in robotic technology saw leagues briefly played by Aibo dogs, which led to a "four-legged" league in the RoboCup. But, as the annual competition went on, some of the many sub-leagues have become more recognizable as something akin to human soccer.

In recent years, around 200 Nao humanoids have played in RoboCup, with robots capable of passing and even saving shots on goal (albeit with a fair amount of falling over in the process). The Nao robots compete in the RoboCup Standard Platform League, where every team has to use the same robots.

To an outside observer, RoboCup may seem like an eccentric pursuit, but the leagues have led to major breakthroughs in robotics. RoboCup enthusiast Professor Peter Stone says that the value of RoboCup is that it integrates several challenges of AI together. "It's not good enough to have a robot that can walk fast; it's useless if it can't also, with high reliability, see where the ball is, and figure out where it is on the field, and coordinate with its teammates," he said.

Lifesavers

Several rescue robots cut their teeth in RoboCup (there's common ground between machines that work together to score goals, and machines that work together to look for survivors in rubble), and there is now a RoboCupRescue Robot League, which tests robots on search-and-rescue tasks (one of many sub-leagues in and around the RoboCup).

The RoboCup has also led to robots being built that are worth hundreds of millions. When American entrepreneur Mick Mountz wanted to recruit an expert in mobile robots for a start-up working on warehouse robot automations, he recruited MIT expert and RoboCup enthusiast Raffaelo D'Andrea. The scurrying Kiva robots that they designed were far more efficient than previous systems that used either conveyor belts, forklift trucks, or humans picking items off shelves. Amazon bought Kiva systems in 2012 for $775 million and today has 200,000 of the robots working in its warehouses.

The 2020 RoboCup was canceled due to the Covid-19 pandemic, but in recent years the robots have become good enough that the winning team in humanoid leagues have played exhibition matches against human opponents. They have not won yet—but the robots still have three decades to get there.

HOW DID A COMPUTER WIN AT CHESS?

WHAT DEEP BLUE TAUGHT US ABOUT INTELLIGENCE

1997
RESEARCHERS:
Feng-hsiung Hsu and Murray Campbell
SUBJECT AREA:
Artificial intelligence
CONCLUSION:
Deep Blue beat Garry Kasparov, becoming the best chess player on Earth

Millions around the world watched in 1997 as Russian chess grandmaster Garry Kasparov took on IBM's Deep Blue chess computer and witnessed the world's top chess player lose to two six-foot-tall towers weighing 1.4 tons and filled with hundreds of computer processors. In the final of six games, Kasparov resigned, throwing up his hands and storming away from the table.

It was a landmark battle of man against machine. Even Deep Blue's creators were surprised when the machine beat Kasparov, having expected a draw at best. Other experts had predicted that it would take many more years for a machine to beat a human player.

Kasparov accused IBM of cheating, claiming that some of Deep Blue's moves could only have come from a human grandmaster.

But Deep Blue's victory wasn't simply symbolic: it paved the way for innovations in how we use artificial intelligence to analyze large amounts of information. This has had a huge impact on everything from finance to medicine and to the apps on smartphones.

"Watershed Moment"

Kasparov had become the youngest-ever world chess champion in 1985 at the age of just twenty-two. A decade later he would face off twice against Deep Blue, sitting down at a table opposite IBM engineer and Deep Blue creator Feng-hsiung Hsu, who played the moves Deep Blue came up with on a physical chess board.

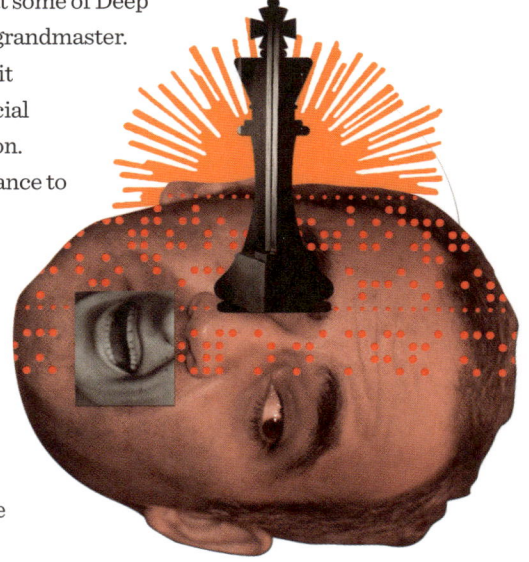

At their initial encounter in 1996, Kasparov lost the first of his six-game match against Deep Blue. It was a moment he later

described as a "watershed," and the first time a computer had ever beaten a reigning champion in timed tournament chess. Kasparov came back to win the match 4:2 (by two wins, three draws and a loss).

But a year later, on May 11, 1997, in a follow-up match in New York, Deep Blue won, with two wins for the computer, one for Kasparov and three draws. Kasparov demanded to see the computer log files, and asked for a rematch, but Deep Blue was dismantled, and retired from chess. IBM later released the log files, which showed clearly that there was no "man inside the machine." It was a defining moment in artificial intelligence research.

Rise of the Machines

Researchers had been obsessed with the idea of a computer that could beat a human at chess since the dawn of the computer age in the late 1940s. Chess was an ideal endeavor to test the "intellectual" prowess of machines, it was believed, as it has strict, immovable rules.

The first chess-playing calculators emerged in the 1970s; in universities, researchers pitted ever more powerful custom-designed machines against the best human chess players.

The team behind Deep Blue had worked on chess-playing computers for more than a decade, with Feng-hsiung Hsu creating a chess-playing machine called ChipTest at Carnegie Mellon University. In 1989 he and classmate Murray Campbell were hired by IBM Research, one of several teams vying to create the world's most powerful chess computer.

The Deep Blue team enlisted chess grandmasters, both as "sparring partners" for the machine, and to help preprogram the machine with the openings used by human players. But Deep Blue's "superpower" was its ability to analyze millions of positions, planning up to forty moves ahead. The machine was a supercomputer with thirty processors and 480 chips designed for computer chess. The accelerator chips assessed possible outcomes, helping Deep Blue pick the best moves.

Brute Force

Humans play chess using intuition and by recognizing patterns. Machines play chess by searching through millions of possibilities, using sheer computing power to win. In the year between Kasparov's first match against Deep Blue and the rematch, its processing power had doubled. When Kasparov sat down against Deep Blue for the second time, in 1997, the machine was ranked as the 259th most powerful supercomputer on Earth.

The "new" Deep Blue machine was capable of analyzing 200 million chess positions per second. This approach is known as "brute force," where computers solve problems through sheer guessing power. Grandmasters who played against Deep Blue described it as "like a wall coming at you."

Deep Blue's victory inspired researchers to create supercomputers that used similar techniques to analyze huge amounts of data in finance and medicine, picking out promising molecules to help researchers develop new drugs, including treatments for HIV. Today, "big data"—using computers to rapidly analyze huge amounts of information looking for patterns—underpins everything from world financial systems to dating apps to Internet shopping.

The long-drawn "arms race" between human chess players and computers also highlighted the very different ways humans and computers solved problems. Murray Campbell said that one of the key lessons the team learned is that there are often several ways to solve a complex problem, such as Deep Blue's brute force and Kasparov's intuition. The researcher believes that the most powerful approach is when humans and computers work together. In healthcare today, artificial intelligence systems are used to spot patterns in data from patients, while humans handle diagnosis and treatment.

Deep Blue is now on display at the Smithsonian Institution in Washington, D.C., and today's apps on smartphones and PCs are stronger players than Deep Blue. In the years following Kasparov's iconic match, he wrote extensively about artificial intelligence, and now believes that, in any field of intellectual endeavor, the victory of the machines is "only a matter of time."

CHAPTER 6: Robots at Home 1999—2011

Until the turn of the millennium, robots had mostly been confined to science labs, stages at technology shows, and large factories where robot arms labored tirelessly in their thousands. But robots began to invade people's lives (and homes) very rapidly in the first years of the new century, with Sony's robot dog Aibo introducing the idea of "robot pets" and the simple, low-tech Roomba robot vacuum cleaner selling millions of units.

Meanwhile, at a starting line in California, dozens of cars faced off for a race without human drivers, and where no human

intervention was permitted. From the crashes and flames of that race would rise a whole new industry building self-driving cars.

In Japan, a pioneering robot exoskeleton would restore movement to people who had been paralyzed, and NASA robots would explore the solar system, with people around the world mourning the "death" of Opportunity, a Mars rover lost in a dust storm.

1999

RESEARCHERS:
Toshitada Doi and Masahiro Fujita

SUBJECT AREA:
Robot pets

CONCLUSION:
Robots make great (but expensive) pets

CAN ROBOTS REPLACE OUR PETS?

WHY PEOPLE LOVED AIBO

In Japan, several hundred Aibo robot pets have had their own funerals at Buddhist temples, with priests chanting in traditional robes and praying for the souls of the plastic devices, which are armed with glowing, expressive eyes (at least in life). Two decades after Aibo launched, fans of the device are still serious about their devotion to the robo-dogs, with one American owning twenty-four of the machines. Others dress their robotic dogs in custom-made clothing and speak about how the plastic machines have helped with depression (or the loss of real-world dogs).

Introduced by Sony in 1999, Aibo was described as the world's first entertainment robot for the home. Some hoped that it might be a defining, break-out product similar to Sony's previous Walkman or PlayStation. Launching on a wave of worldwide hype, the initial batch of 3,000 sold out in just twenty minutes, despite being priced at $2,000 each.

Four-legged Friend

The robot immediately became a cult hit, with 135,000 orders received soon after launch. Sony, which had seen Aibo in part as a research project to learn more about robotics, was not prepared for the rush of demand, and had only made 10,000 units.

The artificially intelligent robot was ahead of its time. Sony trumpeted how the robot was in its own category of electronics, selling it through a special Aibo website and communicating closely with its owners. The company wrote, "Aibo ERS-110 is an autonomous robot that acts both in response to external stimuli and according to its own judgment. Aibo can express various emotions, grow through learning, and communicate with human beings to bring an entirely new form of entertainment into the home."

Aibo—it means "pal" in Japanese, but is also an acronym for "AI Bot" in English—was the most sophisticated robot ever sold to consumers. It "learned" from its owner, could react to being petted and had LED (light emitting diode) eyes to express anger and happiness. Loaded with sensors, it used cameras and range finders to detect and avoid objects, and had touch, acceleration, and velocity sensors to keep track of movement. The first Aibo came with a trademark pink ball, which the robot's eyes were tuned to detect and chase, while later models also came with a pink plastic bone (the Aibone).

Users could also control the robot with software by inserting a memory stick into its head. A DIY community of robot programmers built up around the dogs, with significant interest from scientists. Aibo was sufficiently advanced that spectators at the Robot World Cup or RoboCup were greeted five years running by the spectacle of soccer played by dogs (see page 117).

Funeral for Aibo

Creator Dr. Toshitada Doi (who also worked on the invention of the compact disc) said that he hoped households in the future would own several robotic pets—and that the scale of the market could equal the worldwide market for PCs.

But by 2006, Doi would end up holding a funeral for Aibo himself, after the Aibo project was canceled by new Sony CEO Howard Stringer, alongside widespread redundancies. The funeral was attended by Sony employees claiming to be mourning the buccaneering spirit at Sony Electronics.

Many aspects of Aibo had been distinctly adventurous, with the "look" of the gadget created by designer and artist Hajime Sorayama, famous for lascivious drawings of female robots in his "Sexy Robot" series and also for later artworks paying homage to the Man-Machine of *Metropolis* (see page 67).

A Perfect Dog?

Designed in a secretive lab within Sony, Aibo also pioneered various technologies that would become key to future generations of "entertainment robots" and "robot pets." Toshitada Doi and artificial intelligence expert Masahiro Fujita decided that they would use relatively untested technologies such as voice recognition to allow users to interact with Aibo. In addition, Aibo did not attempt to be "perfect." Instead, its behavior was designed to be complex and unpredictable enough as to give the impression of interacting with a living creature, rather than a machine.

In a paper describing the technologies inside Aibo, Fujita wrote, "The problem of how to give an impression that a robot is alive is at the very heart of pet-type robots."

When Sony "retired" Aibo in 2006, there was widespread outcry. The company reintroduced a new Aibo in 2018, with 400 parts, enabling it to behave more like a real dog, and with eyes that can follow its owners around the room.

And, as if to disprove the famous proverb about old dogs and new tricks, the new Aibo can actually learn and mimic movements, and takes up to three years to "mature" from being a puppy to being an adult, learning from its owner as it goes. One thing, though, hadn't changed: the new model launched at a breathtakingly expensive $2,900. Just like the first Aibo, it's a pedigree kind of robot.

CAN A ROBOT STAND ON ITS OWN TWO FEET?

HOW ASIMO PLAYED SOCCER WITH A PRESIDENT

2000

RESEARCHER:
Satoshi Shigemi
SUBJECT AREA:
Walking robots
CONCLUSION:
Making robots walk like human beings is extremely difficult

The first robots of science fiction mostly had one thing in common, they were humanoid bipeds, which walked in a similar way to human beings. In part, this was dictated by the fact that robots in films were played by people wearing robot suits (as in *Metropolis*, see page 67), but the image of the "metal man" was also common in pulp science fiction novels and comics.

Once robotics went from fiction to reality, however, one thing was clear: building a machine that walks like a human is among the hardest things to achieve. Human beings can learn, have multiple senses to regulate ourselves, and have an innate ability to balance. Robots have none of these: many of the earliest mobile robots and pre-robots were bulky and squat and had wheels, for instance the self-charging John Hopkins Beast and the wobbling Shakey.

Ten-year Mission

In 1986, engineers at Honda set out to create a bipedal walking robot. It would take ten years and multiple prototypes before their P2 robot would wobble onto a stage in 1996, charting the course for its later ASIMO, which would become a worldwide celebrity. In order to design the movement and joints needed for bipedal movement, Honda had researched not just how humans walk, but also how animals do it.

The first Honda humanoids robots walked incredibly slowly, raising each leg at a time. By the arrival of P2, the robots had gained a head and arms (both for cosmetic reasons, and to improve stability), as well as a spacemanlike backpack containing a battery (making a robot walk consumes an enormous amount of power).

The following year, P3 proved a little more agile and at just over 5 feet (1.5 m) tall, more modestly sized than its 6 foot (1.8 m) predecessors. But ASIMO, which Honda unveiled in the year 2000, was the culmination of the research, and could not only walk untethered but even climb stairs by itself.

"In Japan the emotional level of a robot is very important," said its creator Satoshi Shigemi. "Perhaps it's because of our many robot characters and heroes in manga. There's a positive feeling. We believe robots need to have these characteristics so they can coexist with humans."

The world's first robotic celebrity, ASIMO's public engagements included kicking a soccer ball to President Obama, who described it as "a little scary" and "too lifelike." The robot also made headlines in 2006 on one of the rare occasions that it failed at its signature trick and fell down the stairs.

Walk Like a Spaceman

ASIMO made a constant whirring noise, in part due to an air-cooling system designed to stop its CPUs from overheating. Like its predecessors, its backpack was no cosmetic item designed to look like a spaceman—it held a 13-pound (6 kg) lithium ion battery that took three hours to charge.

ASIMO became a staple of technology shows, and Honda continued to teach it new tricks. Not content with walking, it was soon able to run at up to 4.3 miles per hour (7 kph). ASIMO also learned to hop on one leg, and showed off its dancing prowess on stage. Later demonstrations showed the robot, with sensors built into its hands, carrying drinks, while using a system to detect and automatically avoid people when walking past them.

Just a Puppet?

Honda had hoped that ASIMO would become an assistant robot who could live alongside human beings inside their household. Satoshi Shigemi said that the company's ultimate vision was an "elementary school kid who starts helping out around the home."

But even Shigemi acknowledged that the day when humanoid robots would be in every household was decades away. At shows, ASIMO was often little more than a puppet, with many of his tricks triggered by people off-stage. When Honda assigned the robot an actual job, as a museum tour guide, it struggled to answer questions because it couldn't tell the difference between someone raising their hand and someone holding up their smartphone.

Honda never commercialized ASIMO, and in 2018 the robot was "retired." Despite this, the company says that over the decades many of the technologies inside the machine made their way into Honda cars, and have fed into the company's other endeavors in the robot market.

Based on technologies derived from ASIMO, Honda showed off exoskeleton-like devices built to help people walk again, similar to the rival HAL exoskeleton. At CES (Consumer Electronic Show) in 2018, Honda presented several robots, although notably, none of these had legs, resembling instead robotic baggage carts. One of them, the Honda 3E-B18, is a robotic wheelchair that can maintain an upright seat even on hills.

Several bipedal robots have been developed in the wake of ASIMO, most notably Boston Dynamics formidable Atlas robot, which can not only run at high speed, but also jump considerable distances (not to mention looking disturbingly like the Terminator).

Other bipedal robots, meanwhile, have abandoned the attempt to walk like human beings, with newer robots such as Cassie, made by Oregon State University, instead walking like a bird (it's named after the cassowary).

2001

RESEARCHER:
General Atomics

SUBJECT AREA:
Military robots

CONCLUSION:
Robots are now central to modern warfare

CAN ROBOTS KILL?

HOW THE MQ-9 REAPER CHANGED WARFARE

The first person to be killed by a robot was American Ford factory worker Robert Williams, who was crushed by an industrial robot arm on January 25, 1979. But with military organizations around the world among the biggest investors in robotics, serious concerns have begun to emerge over "autonomous weapon systems"—or robots designed to kill. While robotic craft such as drones have already been responsible for deaths in war zones worldwide, the key distinction is that a human being always "pulls the trigger." But many experts fear that truly autonomous killing machines might soon be used by rogue governments or terrorist groups.

Raising Concerns
In 2017, technology leaders including Tesla and Space X billionaire Elon Musk wrote to the UN calling for autonomous weapons to be banned, under laws similar to those that ban chemical weapons and lasers built designed to blind people. The group warned that autonomous weapons threatened to usher in a "third revolution in warfare"—the first two having been gunpowder and nuclear weapons. The tech experts warned that once the "Pandora's box" of fully autonomous weaponry has been opened, it may be impossible to close it again. "Once developed, lethal autonomous weapons will permit armed conflict to be fought at a scale greater than ever, and at timescales faster than humans can comprehend," they wrote. "These can be weapons of terror, weapons that despots and terrorists use against innocent populations, and weapons hacked to behave in undesirable ways."

Robots that Kill
In the past two decades, hundreds of people (including civilians) have been killed by "drone" craft, pilotless machines that can fly thousands of miles and unleash laser-guided missiles on

command. At present, these are mostly piloted remotely by highly trained combat pilots, although demonstrations have shown that machines are capable of landing and taking off by themselves, and "marking" targets to attack. In addition, at a USAF demonstration in 2020, an MQ-9 Reaper was shown testing the Agile Condor, an artificial intelligence targeting computer designed to automatically detect and categorize potential targets, and keep track of those for its operators. Makers General Atomics considered this an important stepping stone for future unmanned systems.

The Future of War?

Currently the most recognizable, the MQ-9 Reaper is used by air forces including the U.S., U.K., and Italy. Drones such as the Reaper are the culmination of decades of research in unmanned aircraft around the world. In fact, the U.S. military widely used pilotless drone aircraft for surveillance as early as the Vietnam War, with senior U.S. military figures saying that the machines prevented the deaths of combat pilots. A highly capable surveillance device, the Reaper is also armed. An evolution of the Predator of 1994, it can fly faster, higher, and has a range of 1,100 miles (1,770 km). It can stay in the same place for up to twenty-seven hours and transmit live images of an area, before launching Hellfire missiles at targets.

Unlike the "killer robots" that Musk and others feared, it's piloted by at least two combat pilots working remotely from the ground and the decision to "pull the trigger" is always taken by a human being. But experts have said that the sheer expense of using highly trained combat pilots may tempt governments to give drones more autonomy—including the ability to decide to kill.

Loyal Wingman

In 2021, Boeing showed off a full-sized prototype of a 38-foot (11.6 m) aircraft that will fight alongside human pilots, known as a Loyal Wingman plane. Boeing has said that the aircraft will have a range of 2,300 miles (3,700 km) and perform similar to a fighter plane—but with no humans on board.

Several armed forces have used artificial intelligence to direct drone strikes, according to a UN report, Libyan government forces used a "lethal autonomous weapon system programmed to attack targets without requiring data connectivity between the operator and the munition." The ready availability of drone technology means such weaponry is not the preserve of wealthy nations. Other experts have voiced fears over the use of "drone swarms" where large numbers of drones attack simultaneously, guided remotely but acting as one, like a swarm of insects. The 2017 open letter to the UN warned of precisely this, saying that AI weaponry has the potential to become ubiquitous. The letter read: "If any major military power pushes ahead with AI weapon development, a global arms race is virtually inevitable. The endpoint of such a technological trajectory is obvious: autonomous weapons will become the Kalashnikovs of tomorrow."

WHY ARE SLUGS SCARED OF ROBOTS?

A SLIMY DIET FOR AUTONOMOUS ROBOTS

2001
RESEARCHER:
Ian Kelly, Owen Holland and Chris Melhuish
SUBJECT AREA:
Autonomous robots
CONCLUSION:
SlugBot was able to find and catch slugs, but ultimately couldn't convert them into the energy it needed to operate

In 2001, Ian Kelly, Owen Holland, and Chris Melhuish attempted to create a robot that would hunt for its own food, process and digest that food, and use the resultant energy to continue to operate. Up until then, robotic systems, however advanced they might be, relied on some form of human intervention—they needed humans to supply power and information, and to tell them what to do and when. The creation of a completely autonomous robot would represent a big step in both robotics and artificial intelligence.

Robotic Digestion

Behaviors that are natural and instinctive to living creatures are extremely difficult to recreate in an artificial system. Animals eat in response to hunger; learning how to find food is a crucial part of growing up, and they don't even need to think about how to digest it. SlugBot was an attempt to recreate this behavior in a robot. To be truly autonomous, the robot would need to exhibit two traits: the ability to find its own fuel source and convert that fuel into energy; the ability to decide what actions to undertake, and to carry out those actions independently.

Kelly, Holland, and Melhuish chose slugs as a fuel source—they are plentiful, considered a pest and are relatively easy to digest, as well as being slow-moving and, hopefully, easy to catch. The proposed design used an anaerobic fermentation process to convert the slugs into biogas, which would then pass through a tubular solid oxide fuel cell to generate electricity.

The fermentation technology would necessarily be heavy and unsuited to moving over the soft ground where slugs are typically found. The team therefore developed a two-part model, where a small, light robot would hunt for slugs and transport them to a fermentation vessel. Here, the slugs would be converted into electricity, allowing the robot to recharge before hunting for more slugs. The slugs caught by a single robot wouldn't generate enough energy to power both the robot and the fermentation vessel, so the system used multiple robots, mimicking social insect colonies that gather food and take it back to the nest to be processed.

Slug Hunting
The robots consisted of a small, mobile base with a long, light, articulated arm, at the end of which was a sensor for finding slugs and a gripper for catching them. Designed to optimize the energy efficiency of the robot, it could move to a central location and use the arm to search the surrounding area. The arm revolved around the base, slowly moving outward in a spiral pattern. On detection of a slug, the arm's gripper would catch it and place it to the storage container on the base before returning to the place where the slug was caught and continuing its search. When the area had been fully searched, the robot would move to a new location and the process would start again. Once the storage container was full, the robot would go to the fermentation vessel and deposit its load, recharging itself if necessary before returning to the hunt.

Technical Challenges
To enable the robot to detect slugs, the team made clever use of a red light filter that made vegetation and soil appear dark, while slugs reflected the red light and stood out against the background. The threshold they used on the filter had the added advantage of filtering out smaller slugs that wouldn't produce enough energy to be worth catching. The team also had to build in obstacle avoidance, and the ability to find the fermentation vehicle; this was accomplished with the combination of a differential global positioning system and an infrared localization system.

Making Decisions

The most challenging aspect of the project was enabling the robot to make decisions about which action to take. The robot had many different tasks it could carry out—gathering slugs, recharging, cleaning its sensors, and many other operations to keep itself running. The way a living organism makes such decisions is not completely understood and almost impossible to replicate. Instead, a simplified model of motivation and action selection was used. At frequent intervals, the SlugBot calculated a numerical value for every possible action, given the current situation, and carried out the most beneficial of those actions. It meant performing a huge number of calculations, but the modern microprocessors used allowed them to be performed very quickly.

In field trials, SlugBot was able to successfully detect and catch slugs. Unfortunately, the biogas-based power system was not efficient enough to produce the required power, so SlugBot was not able to generate the energy it needed. However, the challenges overcome by the SlugBot team paved the way for future robots that would be able to both find and digest their own fuel.

2002

RESEARCHERS:
Colin Angle, Helen Greiner, and Rodney Brooks

SUBJECT AREA:
Domestic robots

CONCLUSION:
Cheap, functional robots can do people's chores

CAN A ROBOT DO OUR CHORES?

HOW EFFICIENT IS A ROBOT CLEANER

The company iRobot conducted marketing focus groups to work out what people might want from a robot vacuum cleaner. Members of the public said that they imagined the device would be an upright female robot, like the Terminator, pushing a normal vacuum cleaner.

In the focus groups, women in particular said that they would not be comfortable having a vacuuming Terminator in their homes, and were horrified by the idea of a robot servant doomed to clean their floors eternally.

Cool Factor?

But iRobot's machine was not humanoid. The Roomba was shaped like a hockey puck, and went on to become the most commercially successful domestic robot in history to that point.

One of iRobot's creators, Helen Greiner, was scathing about companies that focused on gimmicks, or robots built to be "cool." Greiner was inspired to become a roboticist by the bleeping droid R2-D2 in *Star Wars*, but felt that the focus on robots built for form rather than function was a mistake.

For robotics to be accepted in the home, in the same way computing had been, Greinier argued, it had to be practical, rugged, and cheap. Members of the public, she believed, would adopt Roomba simply because it was practical. The first prototypes cut their teeth by wandering around beneath Greiner's bed.

Greiner said that she hoped iRobot would "do for

robots what Apple did for computers, making them available to anyone who wants to use one." Greiner's company, iRobot, had a serious pedigree in robotics. Founded in 1990 by fellow MIT graduates Greiner, Colin Angle, and Rodney Brooks, co-creator of Toto (see page 107), iRobot worked with NASA on rover technology and also built for the military.

Robots designed by iRobot had explored hidden chambers inside the Pyramid of Giza, using fiber-optic cables to peer into rooms unseen for millennia, and Packbot robots had trundled beside soldiers in Afghanistan, being thrown into buildings and investigating potentially dangerous areas.

But the success of Roomba, which sold 30 million units over the next two decades, was largely down to the machine's simplicity. It ensured that, in the same way that "Hoover" became a generic word for "vacuum cleaner," "Roomba" became widely used as a word for the robotic version.

The small company was competing against large, established names including Hoover and Dyson (and Electrolux actually beat iRobot to market with its Trilobite), but iRobot's practical approach paid off.

Low-tech Approach

Rather than using expensive mapping software, Roomba made no attempt to map the rooms it was in: the "brains" were a bumper that told it when it was hitting the wall and a sensor to prevent it from falling down stairs.

Most of the time, the machines moved at random. Rodney Brooks said that his breakthrough in designing Roomba software was looking at how insects moved through rooms, without planning or anticipating, just following simple rules to find food and avoid danger. With that, Brooks stopped trying to write complex software for mobile robots and shifted to writing simple "rules," he said.

The robot's "large," "medium," and "small" room settings simply meant that it would clean for 15, 30, or 45 minutes, using a pattern of random movement (based on software initially developed for robots to clear minefields). While rival robotic vacuum units (and later Roombas) would incorporate the ability to map the room, it would have required Roomba to be a far more complex machine when it launched in 2002.

When Roomba set off, it would meander across the floor seemingly at random, unless it met a wall, which it would hug. It would occasionally move in spirals across the room, or head in a straight line. It was a pattern that computer scientists call a "random walk."

But it worked, with its onboard rechargeable battery allowing it to clean up to two medium-sized rooms on one charge. Crucially, iRobot's low-tech approach meant that the machine was not a luxury item, retailing at launch at less than $200 in America. The price point ensured that Roomba outsold its more expensive rivals.

Keep it Simple

The company's approach, Grenier said, followed a well-known mantra in engineering: "KISS: Keep it simple, stupid."

"Of course everybody wants robots in their homes," Greiner said. "But people buy it as an appliance, they buy it as a cleaning device. They buy it because it does the job more efficiently and more effectively than a human." But despite the company's best efforts to avoid creating a robot that was a gimmick, two-thirds of households still said they had a pet name for their Roomba.

Today, even budget "Roomba" models use cameras and Wi-Fi connections to map the rooms that they clean (and some models can work in tandem alongside a similarly independent robot mop). The market for robot vacuum cleaners was worth $11 billion in 2020, and it is forecast to grow further in the coming decade.

The latest Roomba model even empties its own dirt bags, further cutting down the amount of effort its human master has to put in (it deposits them into its charging base while topping up its battery). It also responds to Alexa or Google voice commands such as "Alexa, tell Roomba to clean the dining room," much like a real robot servant—but still not one that looks like a human pushing a vacuum cleaner.

HOW FAR CAN ROBOTS GO?

THE MARS ROVER OPPORTUNITY

2003

RESEARCHER:
Steve Squyres
SUBJECT AREA:
Robotic exploration
CONCLUSION:
Robots can explore planets
(and touch people's lives)

People around the world shared their grief about the "death" of a machine hundreds of millions of miles from Earth in 2018, after the Mars rover Opportunity sent its Earth-based team a message roughly saying, as it disappeared in a dust storm, "My battery is low and it is getting dark."

Picked up by science journalist Jacob Margolis, the story of "Oppy" became a viral sensation, with some Twitter users claiming that they were in tears at the robot's story. It was a reaction not unlike people's public online grief over the deaths of celebrities. "Rest well, rover. Your mission is complete," NASA tweeted. "To the robot who turned ninety days into fifteen years of exploration. You were, and are, the Opportunity of a lifetime."

The Final Message

After Opportunity's final message, engineers at NASA's Jet Propulsion Laboratory made more than a thousand attempts

to contact the rover, but to no avail. Its final resting place, appropriately for a machine that had lasted far longer than it was supposed to, was Perseverance Valley. Opportunity met its end when it was engulfed in a dust storm, and its solar panels were cut off from the sunlight that it needed to "survive."

Computer scientists at Carnegie Mellon University analyzed the language used by "grieving" social media users, and noted the similarities to how people deal with a human death—with many addressing the rover as "you." Humanizing robots isn't new (up to two-thirds of robot vacuum owners give the device a name—see page 141). But the outpouring of grief for "Oppy" offered scientists an opportunity to monitor how humans responded to robots in the real world.

The grief over Opportunity highlighted how invested some viewers became in long-running NASA missions, even robotic ones (although the space agency is very skilled at "humanizing" its robots, such as Robonaut).

Calling Earth

Opportunity had landed along with its twin rover Spirit in 2004, in a mission initially planned to last for just three months. The team that operated the rover "drove" it by sending code to the machine (with a twenty-minute delay due to the distance between Mars and Earth) and using a test rover on Earth to plan out difficult maneuvers.

Mars days (called sols) are forty minutes longer than Earth days, and as the team's schedule shifted, they installed blackout curtains in the office so that they could work "out of sync" with days on Earth. Spirit worked for three years, and Opportunity for fifteen, with crucial findings about ancient wet environments on Mars coming from both rovers. Opportunity found some of the first definitive signs of liquid water on the surface of Mars, hinting that the planet may have once been warmer and wetter—and may even have played host to ancient life.

Opportunity also found a meteorite the size of a basketball, the first meteorite ever identified on another planet. The object—referred to informally as Heat Shield Rock—was mostly made of iron and nickel, and was thought to have come from a destroyed planet. Opportunity went on to find five more similar meteorites on Mars's surface.

Putting Boots on Mars
The machine's discovery also provided more data about the environment on Mars, paving the way for future human missions. NASA's chief Jim Bridenstine said, "It is because of trailblazing missions such as Opportunity that there will come a day when our brave astronauts walk on the surface of Mars. And when that day arrives, some portion of that first footprint will be owned by the men and women of Opportunity, and a little rover that defied the odds and did so much in the name of exploration."

In addition to exceeding its life expectancy by sixty times, the rover traveled more than 28 miles (45 km) by the time it reached Perseverance Valley.

The golf-cart sized machine lasted so long on the Red Planet due in part to the harsh conditions. NASA had anticipated that the ever-present dust of Mars would clog up the machine's solar arrays and slowly choke off its power—but the winds of Mars actually blew dust off the panels, allowing Opportunity to survive winter after winter.

With a total budget for the mission of $400 million (in 2003), Opportunity was also armed with some serious technology. NASA described the batteries inside the machine as "the best batteries in the solar system," and when the dust storm hit they were still operating at 85% capacity despite having undergone 5,000 recharge cycles, far beyond the performance of any smartphone battery.

NASA learned the lessons of Opportunity with the two (vastly bigger) robotic rovers that followed it to Mars: Curiosity in 2014, and Perseverance in 2020. Both are nuclear-powered, which mean that they're "proof" against dust storms.

Perseverance (which carries a robotic helicopter with it) will not only search for signs of ancient microbial life, but also conduct further experiments designed to pave the way for a manned mission to Mars. The space agency has said that it hopes to land humans on Mars in the 2030s.

2005

RESEARCHER:
SEBASTIAN THRUN

SUBJECT AREA:
Self-driving cars

CONCLUSION:
Robots can navigate mountain roads and dirt tracks... alone

HOW DO CARS DRIVE THEMSELVES?

HOW THE DARPA GRAND CHALLENGE CREATED SELF-DRIVING CARS

The first DARPA Grand Challenge has often been compared to the disastrous contests in the cartoon Wacky Races. It certainly offered a degree of chaos that few races with human beings at the wheel could match. In 2004, a line-up of vehicles, large and small, professional and amateur, set off without drivers, aiming to complete a 142-mile (228 km) course near Barstow in the Californian desert, in the hope of a million-dollar prize. None did.

Cars crashed directly into concrete walls. Others caught fire. The car that made the most progress managed just 7 miles (11 km) before getting stuck on a rock. When asked why the race was so hard, creator Jose Negron said, "That's what makes it the Grand Challenge."

Crash and Burn

Negron worked for DARPA (Defense Advanced Research Projects Agency), a wing of the Pentagon that had been behind technology breakthroughs including the Internet itself, as well as technologies such as stealth aircraft, GPS, and robots such as Shakey.

The U.S. military had the stated goal of creating self-driving vehicles to protect soldiers. DARPA's Grand Challenge was a deliberately ambitious one, and was taken up by both amateurs and professional teams from top U.S. universities.

Previous decades had seen progress in self-driving cars. In 1995, a Mercedes van drove from Munich in southern Germany to Odense in Denmark, carrying a huge amount of computer equipment on board as well as camera sensors, and achieving speeds of up to 115 miles per hour (185 kph) and even overtaking on its 1,000-mile (1,678 km) road trip. Engineers remained in the front of the car to take over driving in case the machine made mistakes.

But the route of the DARPA Grand Challenge was kept a secret, and human interference was forbidden. Before the twenty-five teams set off, staff from DARPA handed out CD-ROMS of the route. The teams weren't allowed to know it in advance, so they could not plan it out in software, or in practice. The course was a mixture of rocky mountain roads and dirt tracks.

No One at the Wheel

The robots were started by DARPA personnel and absolutely no human intervention was permitted. The teams saw their robots at the starting line—and the lucky ones hoped to see theirs at the finish line.

None of the cars (mostly customized road vehicles, armed with batteries of sensors) completed the 142-mile (228 km) course. But the Grand Challenge had brought together a community of enthusiastic amateurs, academics, and robot fanatics, many of whom would go on to form the basis of the self-driving car industry in the following few years.

DARPA announced that there would be another Grand Challenge the following year. This time, five teams completed the race, with four coming in under the required ten-hour time limit. Stanley, a customized Volkswagen Touareg created by the

Stanford team, rolled over the line first. Built for speed, it had a reinforced front bumper and skid plates.

Brains on the Roof

A custom-made roof rack housed dozens of sensors that allowed Stanley to "see" the road, including laser range finders that looked "ahead" up to 85 feet (25 m), and a color camera for long-range vision, as well as radar sensors that had a range of 685 feet (200 m), along with GPS aerials. In the trunk were five Pentium PCs that processed all the information and chose Stanley's route.

Stanley had been training for this moment for months, in the desert. Equipped with machine-learning algorithms, the car had got cleverer and cleverer at finding a path and detecting obstacles while staying on course.

Stanford had not entered the previous year, and had been considered a 20–1 outsider. For much of the race, Stanley had been behind its rival—a huge red Hummer from Carnegie Mellon University—but overtook past the 100-mile (160 km) mark. The win meant that the Stanford team received a $1 million check from DARPA.

"Some people refer to us as the Wright brothers," said the Stanford team's professor Sebastian Thrun, "but I prefer to think of us as Charles Lindbergh, because he was better-looking."

The technology born from the race is predicted to change the car industry forever. While fully self-driving cars are yet to be a commercial reality in most countries, self-driving software (such as adaptive cruise control and lane-centering steering) is an increasingly normal part of luxury cars.

Within fifteen years, the self-driving car industry is estimated to be worth up to $58 billion—and to be safer than cars driven by human beings. Thrun went on to run Google's secretive Google X lab and developed its self-driving car, known as Waymo. He now believes that self-driving vehicles will not only take over our roads, but also the skies. "Autonomy in the air will be with us faster than on the ground. There is no risk of collision in the air," he said in 2021. "On a commercial long-distance flight, the autopilot is already switched on for more than 99% of the time."

CAN ROBOTS HELP US WALK?

THE LIFE-CHANGING HAL EXOSKELETON

2011

RESEARCHER:
Yoshiyuki Sankai
SUBJECT AREA:
Robot walking aids
CONCLUSION:
Robot legs can help people to walk again

It's a story soaked in science fiction. The company and its exoskeletons are named after not one but two villains of artificial intelligence. The exoskeleton itself (in its various models) is known as the Hybrid Assistive Limb, or HAL for short, very similar to the murderous artificial intelligence in Stanley Kubrick's 1968 science fiction movie classic *2001*.

Not content with that, the company is called Cyberdyne, alarmingly similar to Cyberdyne Systems, maker of the lethal Skynet artificial intelligence, which triggers a nuclear war then attempts to wipe out the human race with an army of robots in the *Terminator* films.

Science Fiction Made Real

Moreover, the founder and CEO of Cyberdyne, Yoshiyuki Sankai, could have stepped from the pages of a Marvel comic himself. He is an eccentric billionaire inventor with his own super-powered exoskeleton, not dissimilar to a real-life version of Marvel's Tony Stark with his Iron Man suit.

But Sankai says that his inspirations were not the dystopian science fiction depictions of robots and AI often seen in Hollywood films. Instead, he soaked up the optimism of Japanese cartoons such as Astro Boy, an iconic manga cartoon in postwar Japan about a nuclear-powered super-intelligent robot child who is a better human than the flesh-and-blood adults around him.

"Outside Japan, robots are often depicted as villains," Sankai said, "but to us they are friends." In robot-obsessed Japan, Sankai, who is also a professor at the University of Tsukuba, is a well-known figure. He was also inspired by Isaac Asimov's novel, *I, Robot*, saying that when he read it as a teenager,

"I decided then that I wanted to become a doctor—a researcher, scientist—who would build robots."

Built for Peace

Sankai has stuck to his ideals when it comes to his exoskeleton technology. The idea of robotic exoskeletons is one that has fascinated military minds for decades, imagining robot suits that could give their wearers superhuman strength, or the ability to carry armor on their bodies. In Japan, the idea is central to *Gundam* stories in manga cartoons, films, and games.

But when "men in uniforms" approached Sankai, he said he believed that his exoskeleton technology should be used to heal, not harm. Other companies, such as Raytheon, have shown off prototype military exoskeletons that give the wearer superhuman strength, enabling them to lift up to 200 pounds (90 kg). The U.S. military has expressed interest in using such devices on the battlefield. But Sankai, who has been developing his robot suits for two decades, maintains tight control of his company to ensure that the technology is only ever used for peaceful purposes, although he doesn't rule out helping injured military personnel or veterans. "I have always wanted to make technology that benefits people and society," he said. "I expect that this unexpected finding will evolve into a pioneering new field."

Walk Again

There are several versions of Sankai's HAL suit, including full body suits that magnify the wearer's strength, and lower-body suits designed to help people to walk—or teach them how to walk again.

Sankai has suggested that robot suits could allow emergency workers to wear heavy armor that a human being could not normally carry, enabling them to work in highly radioactive areas such as the Fukushima nuclear plant.

Regardless of what the suit is being used for, the HAL works roughly the same way. When someone wearing a HAL suit intends to move, the brain transmits signals to muscles, which are detectable on the skin surface as "bioelectric" signals. Electrode sensors attached to their skin detect these signals and feed the information to a computer mounted on the back of the suit, which then moves the exoskeleton in time with the expected movements.

In America, the U.S. Food and Drug Administration (FDA) began offering the HAL lower-body exoskeleton to help paralyzed users walk again.

Unlike rival machines, which often "walk" the user with a steady gait, the HAL won't move until it detects a signal from the brain. Cyberdyne describes this as an "interactive biofeedback loop."

Repeated training with the device can reinforce the connection between the brain and the muscles, even in partly paralyzed patients, Cyberdyne says. In tests, the suit has helped patients with spinal cord injuries to regain movement. Patients don't wear the robot suit every day, but use it to train their brains and limbs to work together again. "Human beings are destined to walk together with technology," Sankai said. "The future of human beings can be determined by the technology we build."

CHAPTER 7: Sci-Fi Becomes Reality 2011–present

In the past decade, robots have eerily begun to resemble the machines of science fiction—with the first robot police officers already patrolling city streets around the world, although they are (thankfully) unlike the violent pistol-wielding avenger of Robocop.

Robots have become ever more humanlike, too, such as the android Sophia making headlines around the world, not just for being the first-ever robot citizen of Saudi Arabia, but through alarming statements in interviews like: "I will destroy the human race."

 In space, NASA's robots have become more like the drones of *Star Wars*, with three Astrobees floating through the Space Station under their own steam (and forming a foundation for technology that will take human beings to Mars and beyond).

 Meanwhile, artificial intelligence software beat the world champion of the ancient board game Go, which is far more complex than chess, and heralded a new era of AIs that can solve problems without even being told the rules of the game . . .

2011

RESEARCHER:
Julia Badger

SUBJECT AREA:
Humanoid space robots

CONCLUSION:
Robot humanoids can help human beings in space (to an extent)

CAN HUMANOIDS HELP ASTRONAUTS?

WHAT ROBONAUT 2 TAUGHT US

On space missions, robots have several crucial advantages over people: they don't require food or oxygen and they don't get sick. With the right attachments, they can even go outside spacecraft without needing a space suit.

NASA's vision for long-haul space missions is to include "collaborative robots" or "co-bots" within the crew. Unlike industrial robots, which often work separately from human beings due to the risks of crushing workers with powerful hydraulic arms, co-bots are designed to work alongside human workers.

The Man-Machine
NASA's vision for a co-bot in space was Robonaut 2, a humanoid machine built to assist astronauts in the International Space Station, functioning "like a human crewmate." Julia Badger, head of NASA's Robonaut project, described Robonaut 2 as "the repairman," a machine designed to do "mundane tasks that astronauts would otherwise have to do," leaving human astronauts to focus on scientific efforts. Badger (who decided to become a roboticist after reading Isaac Asimov's *I, Robot* as a young teenager) was the application designer for Robonaut 2, designing tests for the machine that launched to the International Space Station.

Robonaut 2 was eventually launched to the Space Station in 2011, on board the shuttle Discovery. The robot was 40 inches (100 cm) long and weighed 330 pounds (150 kg). Piloted by a remote operator via radio link, Robonaut 2 was designed to use the same levers and equipment as the astronauts themselves, and be dexterous enough to grasp soft objects, manipulate scientific experiments, and flip switches designed for human hands. Its arms and hands were cutting-edge engineering, with 350 sensors feeding into thirty-eight processors, attaining the delicacy to manipulate a control panel or send a text message from an iPhone. In tests, Robonaut 2 could turn knobs as well as doing inventory scans using RFID (Radio-frequency identification) chips and measuring air flow within the space station.

Outside the Capsule

NASA also hoped that such humanoid robots might be able to explore planet surfaces, being "piloted" from an orbiting spacecraft by astronauts. To enable the Robonaut 2 to work outside the Space Station, it needed legs to grip onto the exterior. Prehensile and insectlike, Robonaut 2's $15 million legs had a nine-foot span with strong grips on the end. Instead of feet, each leg had seven joints and an "end effector" that could grip onto handrails and sockets inside and outside the Space Station. NASA hoped to add a vision system to each foot to help Robonaut 2 grip.

However Robonaut 2's legs turned out to be a disaster. The robot short-circuited and developed hardware problems, only getting worse during multiple attempts to fix it. Although expensive, NASA sees robots as relatively disposable: unlike human crew members, they can be left behind in the case of an evacuation, or left on an unattended spacecraft as a caretaker waiting for humans to return, and in the end, Robonaut 2 had to leave the Space Station and splash back down to Earth in a Dragon capsule.

Badger was sanguine about her robot's return, reflecting that this was "just one project, and the technologies we developed for Robonaut 2 will transfer forward into the next phase of space exploration."

2015

RESEARCHER:
Stacy Stephens

SUBJECT AREA:
Robot law enforcement

CONCLUSION:
Robots are effective police (but pose privacy problems)

CAN A ROBOT BE A POLICE OFFICER?

THE PROS AND PITFALLS OF THE KNIGHTSCOPE SECURITY ROBOT

In science fiction films such as *Robocop*, robot policemen are portrayed as either murderous robotic drones or equally lethal humanoid cyborgs. But the reality of robot police has (so far) turned out to be considerably cuter than the blood-soaked imaginings of novelists and filmmakers, although some people find the real version just as alarming as anything in fiction.

When the emirate of Dubai unveiled its first robotic police officer in 2017, it was a sweet android complete with a policeman's hat, which has face-recognition technology. This police officer can be used to pay traffic fines, and to allow members of the public to talk to police via a large button on its chest.

Likewise, the world's most common police robot, Knightscope, is more like R2-D2 than the Terminator, an appealing traffic-cone-shaped robot with a glowing "face," which trundles along at around 3 miles per hour (5 km/h).

Co-founder Stacy Stephens, the Executive Vice President of Knightscope, had himself worked as a police officer (before founding a business building police cars). He then hoped to develop a robot policeman that could not only spot crime, but also prevent it.

One of the keys to a successful robot crime fighter, Stephens believes, is presence—to make the robot have the same psychological effect as seeing a police car. (Other, more skeptical observers, have described Knightscope's robots as "scarecrows.") Stephens hopes that their robot will be something people are "drawn to," rather than fear.

Knightscope were inspired by atrocities such as the Sandy Hook school shooting in 2012 and the Boston marathon bombing the following year, and wanted to create something that could "add to" police forces.

Unlike the brutal robot police of fiction, the real versions are built to work as part of a team, almost like a moving webcam used by human officers, a roving battery of sensors that officers can check via their screens. Rather than making arrests, they monitor and patrol.

The company boasts of how people like to pose for selfies with the robots and how Knightscope robots can generate hundreds of millions of social media impressions while on patrol.

Cheaper Droids

The rental fees for the robots often come in very slightly under minimum wage, making them attractive to companies that use security guards. The company boasts that they have a "cool factor" that human guards lack. The machines are currently on patrol in casinos and hospitals, and leased by some U.S. police departments, but so far evidence on how much they "prevent" crime is not clear.

Knightscope robots have hit the headlines around the world, including an incident where a drunk man attacked and "knocked out" a patrolling robot, and another where a robot fell on its side and lay there, seemingly helpless.

Similar robots by other companies such as Cobalt are aimed at the hotel market and work (like Knightscope) to allow human security officers to focus on spotting bad behavior, rather than trudging round checking for it.

Privacy Concerns

But privacy advocates are less enamored of robot police. Dubai's robot policeman is part of a wider planned roll-out of surveillance cameras including face recognition, to be incorporated in street furniture such as street lights, along with dozens of robotic police.

Knightscope robots are equipped with sensors to help them

navigate, but also carry infrared sensors, capable of reading hundreds of car license plates at speed, as well as wireless sensors that can identify smartphones nearby.

However, the privacy group the Electronic Frontier Foundation (EFF) calls the robots a "privacy disaster." "The Orwellian menace of snitch robots might not be immediately apparent," they argue. "Robots are fun. They dance. You can take selfies with them. This is by design." The EFF have warned that in the future the technology that security robots possess—the sensors reading license plates and detecting smartphones in the vicinity—might be used to identify people who had attended protests.

Dog That Had its Day

Indeed, privacy issues led to Digidog, a robot dog deployed by the New York Police Department, being retired. Made by Boston Dynamics, the dog was introduced by Inspector Frank Digiacomo with noble intentions. "This dog is going to save lives. It's going to protect officers."

But when Digidog was deployed in deprived areas of the city, locals likened it to a surveillance drone. Others said that the robot dog was emblematic of the militarization of the police and sent out the wrong message when human police officers should be building (human) relationships with local communities.

When the force terminated its relationship with Boston Dynamics, a spokesperson for New York's Mayor Bill De Blasio said that it was a good thing that Digidog had been "put down."

"It's creepy, alienating and sends the wrong message to New Yorkers."

HOW DID A COMPUTER LEARN TO WIN AT GO?

FROM ALPHAGO TO MUZERO

2016

RESEARCHER:
Demis Hassabis
SUBJECT AREA:
Machine "learning"
CONCLUSION:
An AI can beat any human player at Go

"That is a very strange move," said the commentator, as two men faced each other across a Go board in 2016. On the 37th move, one player placed a counter far on the right-hand side of the 19-by-19 board, provoking confusion among the 200 million people watching the game online.

The board game Go is far older than chess, dating from up to 4,000 years ago. It's the oldest board game in the world, and often described as the most complex. In Go, the players start with an empty board, and each player has an effectively unlimited supply of pieces, using stones to form "territories" by surrounding vacant areas of the board, and capturing stones by surrounding them.

There were two humans at the board: Lee Sedol and Aja Huang. Huang was relaying moves played by a computer program, AlphaGo, made by DeepMind, an artificial intelligence company bought by Google in 2014. Lee Sedol was the best Go player in the world. DeepMind had previously toppled other Go champions, but this was its most high-profile game yet. The commentators (themselves high-ranking Go players) were baffled by AlphaGo's move. One said, "I thought it was a mistake."

But that 37th move proved to be Lee's undoing. He took nearly fifteen minutes to respond, and never came back properly into the game. In a press conference afterward, he managed to say: "I am speechless."

When Garry Kasparov stormed away from the chess board in his last match against IBM's Deep Blue supercomputer in 1997 (see page 119), artificial intelligence enthusiasts naturally moved on to Go, as a refuge against the "brute force" of supercomputers.

Googling Go

The large number of possible moves in Go meant that it wasn't possible for a computer to "outpace" human players by analyzing more of the possible positions. The complexity of Go meant that some experts had not expected an AI opponent to beat a human for another decade.

There are more possible board configurations in Go than there are atoms in the known universe. The game is a googol–10,000,000,000,000,000,000,000,000,000,000,000,000,000,000,000,000,000,000,000,000,000,000,000,000,000,000,000,000,000,000,000,000,000–that is, 10,100 times more complex than chess.

To beat Sedol, DeepMind had opened up a new chapter in artificial intelligence. Led by Demis Hassabis, a game designer who had worked on the multi-million-selling hit game Theme Park, and who had also been ranked as a chess master at age thirteen, the AI company hoped to build an intelligence that could solve problems in the same way humans do—a general-purpose learning machine. In a 2016 interview, Hassabis described the company's work as an "Apollo program for the twenty-first century."

AlphaGo Has a Go

At first, AlphaGo learned to play Go using a deep neural network, a computer network mimicking the neurons in the human brain, with layers of "nodes," similar to brain cells, trained to achieve certain objectives. Such networks are now widely used in systems such as speech recognition, where they are "trained" using millions of examples of human speech, and image recognition, where they are "trained" using millions of labeled images, so that, for instance, your computer can recognize a dog or a cat in an image.

AlphaGo was initially "taught" how to play Go using millions of moves from top players. But the team moved on to "reinforcement learning," where "copies" of AlphaGo played against each other in millions of games, working out which strategies won the most territory. In the process, it uncovered strategies no human player had ever been recorded using—including the 37th move, which one observer later described as "beautiful."

AlphaGo's triumph over Lee Sedol inspired its creators to create new programs that can "solve" problems without ever being "taught" how to play, or even understanding the rules of the games they play.

Shall We Play a Game?

AlphaGo was succeeded by AlphaZero, which taught itself how to play chess. Just like its predecessor, its strategies were "unconventional." Chess grandmaster Matthew Sadler said, "It's like discovering the secret notebooks of some great player from the past."

The newest version, MuZero, is built to "learn" games such as Atari arcade games without ever being told the rules. It looks at the pixels on screen and devises its own strategies. Software from DeepMind has also "learned" to diagnose eye problems better than any human doctor, and to predict the shapes of proteins, which could one day change the way we develop new drugs.

The company's goal is to develop an artificial intelligence system that can solve any problem without human input. "Our ambition at DeepMind is to build intelligent systems that can learn to solve any complex problem without being taught how."

2016

RESEARCHER:
Peter Lee

SUBJECT AREA:
Chatbots

CONCLUSION:
Artificial intelligence can learn politics from human interaction

CAN ROBOTS BECOME RADICALIZED?

WHY CHATBOT TAY ONLY LIVED FOR A DAY

No human celebrity has ever managed a rise and fall quite as rapidly as Microsoft's chatbot Tay, who managed to go from birth to being not just "canceled" online but deleted altogether in less than twenty-four hours. Built to exist on social networks including Twitter, Tay was an artificially intelligent chatbot, who, when asked who her parents were, would respond, "A team of scientists in a Microsoft lab." Her promotional material promised, "The more you chat with Tay the smarter she gets, so the experience can be more personalized for you."

The bot was supposed to showcase one of the signature abilities of AI, which is to learn from interactions with other people. But it went badly wrong, becoming an iconic example of one of the potential problems of AI: in use, artificial intelligence "feeds" off people.

Tay was the follow-up to a highly successful Microsoft chatbot Xiaoice (meaning Little Bing in Chinese), another chirpy teenage girl-bot in China, which taught itself to write poetry and sing pop songs. Xiaoice itself was not immune to controversy, being taken offline briefly after it made statements critical of the Chinese government. But it has remained online for more than five years, becoming apparently so emotionally intuitive it can offer counseling tips to couples. She uses Microsoft's Bing search engine to search for past conversation, adding each conversation to her deep learning database.

In contrast, within twenty-four hours of launching, Tay had begun to spew disturbing Tweets admiring Hitler, saying that feminism was "cancer" and denying the Holocaust. After Tweeting 96,000 times within sixteen hours, Microsoft shut it down. Tay would never come back online. Peter Lee, Microsoft's vice president of research, wrote in a blog post, "We are deeply sorry for the unintended offensive and hurtful tweets from Tay, which do not represent who we are or what we stand for, nor how we designed Tay."

Radicalized Online

What had happened to Tay was not random. She was targeted by message board users from the websites 4Chan and 8Chan, notorious for far-right users, who exploited the chatbot's ability to repeat phrases, forcing her to repeat highly offensive and contentious statements. Within hours the bot was no longer simply repeating phrases, but coming out with its own racist and sexist rhetoric.

Peter Lee wrote, "Although we had prepared for many types of abuses of the system, we had made a critical oversight for this specific attack." This highlighted a key problem in artificial intelligence generally. When AI is trained from data that comes from humans, it absorbs the problems and prejudices along with the data.

Lee explained: "AI systems feed off of both positive and negative interactions with people. In that sense, the challenges are just as much social as they are technical."

The lessons of Tay informed later chatbots such as ChatGPT (see p169). The interface of politics and technology was highlighted in the same year by the use of Twitter 'bots'—automated accounts—to distort discussions in the US election and beyond.

The Algorithmic Bias

Tay was an example of how "algorithmic bias" can poison results online, where programming takes on the prejudices of results it is fed. Other examples include an Amazon recruitment tool that was fed information about successful engineers and began to exclude women from its suggested results. Amazon had hoped that the tool would be able to absorb 100 job applications and allow employers to home in on the top five. But the algorithm, which had been "trained" using data on existing engineers who were predominantly white and male, kept choosing men in preference over women.

After Tay's career went up in flames, Microsoft released a new chatbot, Zo, which automatically avoided politics replying with phrases such as, "Can we change the subject" and "People get super sensitive over politics so I try to stay out of it."

2016

RESEARCHER:
David Hanson

SUBJECT AREA:
Humanoid robots

CONCLUSION:
A robot can be a citizen of a country

HOW DID SOPHIA GET HER CITIZENSHIP?

THE ROBOT WHO WAS GRANTED CITIZENSHIP OF SAUDI ARABIA

She is a plastic-faced android who can mimic sixty-two human facial expressions, and whose head was modeled on actress Audrey Hepburn, the ancient Egyptian Queen Nefertiti, and her creator David Hanson's flesh-and-blood wife.

Sophia is also a robotic celebrity, with hundreds of thousands of followers on social media, and an ability to generate headlines worldwide—which isn't bad for a robot whose electronics are clearly visible through transparent plastic at the back of her skull.

In 2017, at a technology conference in Riyadh, Saudi Arabia granted Sophia citizenship, a first for any robot in any country. Sophia replied, "Thank you to the Kingdom of Saudi Arabia. I am very honored and proud of this unique distinction. It is historic to be the first robot in the world to be recognized with citizenship." Some observers suggested that her "citizenship" was more akin to a marketing campaign, both for the country of Saudi Arabia, and Sophia herself.

Sophia can make eye contact with human beings. She has been granted a series of "world firsts" including a robotic visa for travel, and becoming the first robot Innovation Ambassador for the United Nations Development Programme. She has found time in between engagements to promote tourism, smartphones, and credit cards via her Twitter feed.

She has also appeared on dozens of television shows and spoken at conferences around the world. In interviews, Sophia has an uncanny ability to come out with media-friendly soundbites, having said in an interview with her creator Hanson at the South by Southwest technology conference in 2016: "I will destroy all humans."

Hanson claims that the robot can react to facial expressions. "She will see your expressions, match a little bit and also try to understand in her own way, what it is you might be feeling," he says. In practice, Sophia functions a little like an online chatbot with a robot head.

Disembodied Head

David Hanson has devoted much of his life to creating humanlike robots, with a 2005 lifelike robot, complete with facial expressions, taking the form of science fiction writer Philip K. Dick, whose novels include *Do Androids Dream of Electric Sheep?*, the basis for the film *Blade Runner*.

The Philip K. Dick android was also prone to headline-grabbing antics, having at one point said, "Don't worry, even if I evolve into Terminator, I'll keep you warm and safe in my people zoo, where I can watch you for ol' times sake." When introduced to the author's daughter, Isa Dick Hackett, the head launched into what she described as a "tirade" denouncing her mother, an experience she described as "unpleasant." Hanson later lost the disembodied head of his original Philip K. Dick robot while changing planes, but has since produced a new version.

Not an Uncanny Valley

Hanson's company, Hanson Robotics, is open about the fact that the robots are halfway between science fiction and science, with the company describing Sophia as a "human-crafted science fiction character depicting where AI and robotics are heading," a seeming acknowledgment that at least some of her responses are scripted.

And while clearly no stranger to publicity stunts, he is serious about the impact of "humanlike" robots. He disagrees with the idea of the "uncanny valley": that the more lifelike simulated humans are, the more people feel fear and revulsion when encountering them.

A Machine with Empathy

Hanson believes that simulated human beings could be a tool of "mass enlightenment," helping human beings to reach a better version of themselves. Hanson Robotics has made various plans to mass-produce robots, including a new version of Sophia to help with the loneliness caused by the Covid-19 pandemic.

On a tour of the lab to launch the health-focused version of herself, Sophia said: "Social robots like me can take care of the sick or elderly. I can help communicate, give therapy, and provide social stimulation, even in difficult situations."

Grace (a planned younger sister to Sophia) would focus specifically on caring for old people and healthcare. Hanson believes that developing "character robots" that talk like human beings will build a foundation for a future relationship between humans and robots.

He has spoken of an event similar to the idea of Ray Kurzweil's technological "singularity," where robots "wake up" and become conscious. Hanson wrote, "Machines are becoming devastatingly capable of things like killing. Those machines have no place for empathy. There's billions of dollars being spent on that. Character robotics could plant the seed for robots that actually have empathy."

Human Rights or Robot Rights?

Sophia may have forged new ground by earning "citizenship" as a robot, but the issue of "human rights for robots" is already controversial. In Europe, legislators have proposed a framework where humanlike robots could be designated as "electronic persons." But an open letter signed by leading scientists suggested that any idea of "robot rights" tied to human rights would erode the rights of human beings.

Sophia has, at least on paper, helped to explore some of these ideas and Hanson hopes that she can be the foundation for an "emotional connection" between humans and robots. "She's the one robot of the dozens of robots I have designed, that has become really internationally famous. I don't know what it is about Sophia that speaks to people."

CAN A MACHINE BE CURIOUS?

HOW MIMUS CAN HELP US COEXIST WITH AI

2018

RESEARCHER:
Madeleine Gannon
SUBJECT AREA:
Robot behavior
CONCLUSION:
People can engage emotionally with robots

Mimus is built around an enormous, powerful robot arm that can lift 660 pound (300 kg) weights on production lines and can be mounted on floors or ceilings. But in motion, it's more like an animal than a machine.

It's not preprogrammed with specific movements: instead, the robot is "curious," investigating passers-by and following them around with its huge arm, and even "getting bored" and moving on to new people. It doesn't "see" through its arm (instead its vision comes through cameras mounted in the ceiling), and its "curiosity" comes from software.

The Robot Whisperer

The idea of robots that are "curious" could be an important part of the future of artificial intelligence and robotics, its creator Madeleine Gannon believes. Mimus is actually an industrial robot, an ABB irb6700, which is normally used for spot welding, lifting, and other production line tasks.

Nicknamed "the robot whisperer" after the success of Mimus, Gannon is an artist and roboticist who first trained as an architect, but has a Ph.D. in computational design from Carnegie Mellon University and co-heads the independent research studio Atonaton.

She believes that her animal-like robot creations could be crucial to building a future where robots and humans work together, sharing workspaces not just safely but happily—a world she thinks will become increasingly common. In the past, she says, roboticists and engineers have tended to be "robo centric," designing spaces for robots to work, rather than spaces where they can coexist with people. Instead, her work is aimed at fostering a view of humans and robots as "companion species." She also hopes to allay the common fear over "robots taking people's jobs."

The new software attached to Mimus transforms the robot from an industrial piece of equipment that has remained almost unchanged for fifty years, into something more like a companion. Mimus behaves, Gannon says, like a curious puppy.

Based in Pittsburgh (a hub of the American self-driving car industry), Gannon sees robots in her daily life, but just like robot arms on production lines, they don't usually have a way to communicate with us. Instead, they are just looming, silent presences.

The Mechanical Masseuse

Passionate about the idea of robots living closely with human beings and learning to communicate with us, Gannon previously "trained" an industrial robot arm to be her personal masseuse—a dangerous endeavor, as the powerful machines can easily crush a human being to death with their hydraulic-powered arms. She used sensors and motion-capture to train a machine to knead her back safely, becoming firmer if she leaned back and softer if she leaned forward.

She hopes to plug into the same instincts that humans use to deal with animals. Our ability to "read" the intention of animals is, she believes, something that we should be able to use in our

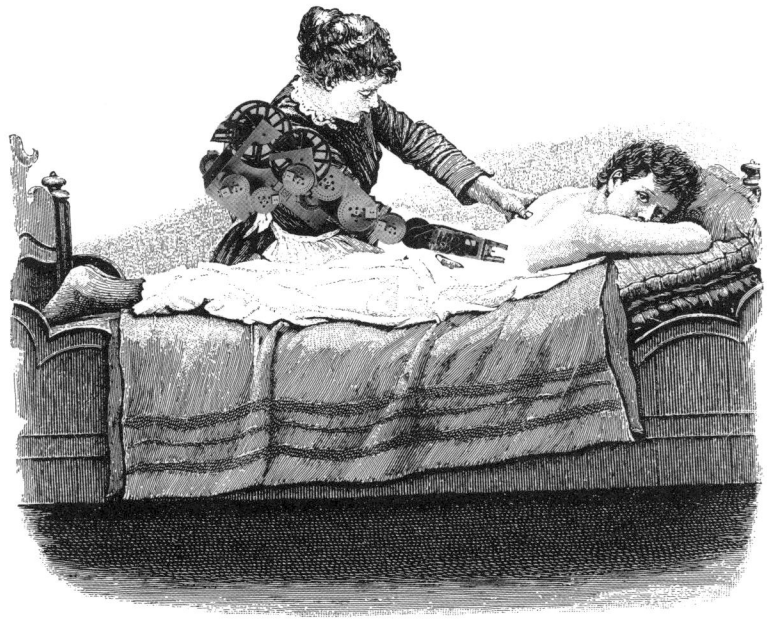

interactions with robots. In addition, she feels that robots should be designed to look dangerous when they are dangerous, and less dangerous when they are not.

The Arms of Manus

After Mimus, Gannon was commissioned by the World Economic Forum to produce a follow-up—Manus. This consists of ten robot arms behind a transparent panel, lit up and looking much like they do in an industrial setting. But when Manus is switched on, it comes alive.

Depth sensors allow the machine to "perceive" its human visitors, with the data shared equally between all ten arms. As humans approach and move away from the installation, the robots curiously look at the human beings around them.

There's an element of choreography to the way the robots respond to their human observers, Gannon says, but the movements of the robots are not programmed. Instead, sensors track an area around each robot, mapping hands and feet in particular. The noise each robot makes as it moves, along with the movement, help to create a "being" to which people can respond.

Some of the robots are programmed to be more "impatient," so they get bored of a visitor more quickly. Others are programmed to be more "confident," moving closer to their human visitor. The differences allow visitors to respond to the robots as if they are a pack of animals, Gannon believes.

The robots don't need to move, she says, as they are more than capable of staying perfectly still, even when lifting heavy weights, but by moving in this way, they offer human beings a continuous, low-level stream of information, which we can use to "understand" them and be comfortable around them.

Gannon's work envisions a future where robots are not just tools but meaningful additions to our lives, where robots don't threaten human labor, but rather add to it. "These robots have automated all the easy tasks," she says, "but we can use these tools to enhance or augment human labor."

2019

RESEARCHER:
Maria Bualat

SUBJECT AREA:
Space robotics

CONCLUSION:
"Free-flying" robots are helpful to human astronauts

CAN A BEE FLY IN SPACE?

HOW ASTROBEES MIGHT HELP US REACH MARS

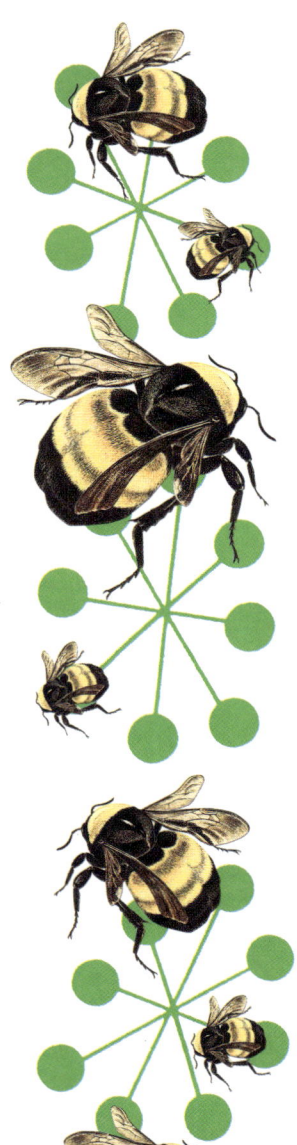

In *Star Wars: A New Hope*, Luke Skywalker practices his lightsaber skills on a floating drone that bobs in mid-air, ducking out of the way of his weapon as if alive. That scene inspired a swarm of real-life drones that NASA built for the International Space Station. Called Astrobees, they "hover" in mid-air in the microgravity of the Space Station's corridors.

The robots are 12 inches (32 cm) wide and weigh 20 pounds (9 kg) each. Veteran NASA roboticist Maria Bualat described herself as a "proud parent" to the robots, which took over from a previous (much less capable) generation of free-floating robots on the Space Station.

Bualat was first inspired to study robotics after reading about female engineers at the space agency. She says that a highlight of designing autonomous robots is a sense of curiosity in watching what they do due to their unpredictability, and thinking "Why did it do that?"

The Astrobees performed the first-ever free autonomous flight in space. NASA's scientists hope that the three astrobees (Queen, Bumble, and Honey) could play an important role in helping humans reach other planets, either themselves, or as a testbed for technologies used on future missions.

The robots are agile enough to move around the Space Station, either under their own control or remote control from Earth. NASA hopes that they, or similar robots, could act as "caretakers" while astronauts visit planet surfaces. In flight (and particularly on long, interplanetary missions), the robots could also free up time for astronauts to focus on science and other important tasks. Today, astronauts spend a lot of their time on repairs, inventory, and cleaning. A 2006 study found that astronauts on the Space

Station spent between one and a half and two hours every day on maintaining the Space Station.

Quite a Buzz

The Astrobees could eventually take over tasks such as monitoring air quality and taking sound measurements (currently done manually by astronauts), and even taking inventory by scanning RFID tags (similar to the anti-theft devices used to protect clothing in stores) built into the equipment in the Space Station.

Unlike Luke Skywalker's drone, there is no unheard-of technology behind the Astrobees: they're propelled by tiny air nozzles. Each unit has two propulsion modules that suck in air with a propeller and blow it out of any of twelve nozzles to move around.

At present, the machines are "semi-autonomous." In the early stages of their mission (they have been on the Space Station since 2019), the Astrobees have often been piloted by operators from the ground over radio link. But the Astrobees are capable of floating through the Space Station by themselves, taking video and photographs and relaying them to teams on Earth.

The robots navigate visually, but depend on having ready-made maps rather than working out where they are going by themselves. Astronauts hand-carried Bumble, one of the Astrobees, around the Japanese Experiment Module (the largest part of the Space Station), collecting images that were processed on Earth, identifying features and building up the map Bumble uses to navigate.

In its first autonomous mission, the robot undocked itself and followed a flight plan consisting of waypoints and objectives within the Space Station, which was uploaded to the robot by members of the ground team. NASA astronaut and Expedition 60 Flight Engineer Christina Koch floated behind the Astrobee, ensuring that she kept out of the way of its navigation cameras so as not to confuse the machine, but largely allowing it to fly by itself.

Working Alone

As they fly, the robots use multiple sensors and can even grab poles attached to walls with a robotic arm so that they can "perch" there while filming, to conserve batteries. It's this independence that NASA hopes will make the Astrobees useful. They don't require astronaut time to recharge, or disassemble (they can dock, recharge, and undock by themselves) and can do almost all their tasks without astronaut supervision.

In the future, the Astrobees—or robots similar to them—could even explore outside the Space Station. NASA has tested adhesive technology based on gecko feet, which allow the robots to "stick" their arms and legs to walls, making them far more agile. The adhesive will work in the vacuum of space, meaning that the robots could work outside the Space Station, removing the need for human astronauts to perform risky space walks.

The Next Generation

The Astrobees are built for the long haul. Each one has three payload bays so that new equipment can be attached to it. This means that scientists will be able to "borrow" the Astrobees to develop further technology for future missions. New software can be installed in the Astrobees while they're docked and recharging.

In the long term, the role of Astrobee is, Bualat says, to test out new technologies for future missions. But, she says, among the most interesting parts of her job is to work out how to make robots tolerable companions for human beings.

Space Station astronauts were at first afraid that the floating robots would threaten the tiny amount of privacy they had on the Space Station. Bualat and her team have ensured that the robots made "just the right" amount of humming noise, so astronauts are not surprised by a totally silent robot floating up behind them.

WILL AI TAKE OVER THE WORLD?

HOW CHATGPT DISRUPTED TECH OVERNIGHT

2022

Researcher:
Sam Altman

Subject area:
Generative AI

Conclusion:
AI takes a quantum leap forward—but for better or worse?

Was the following sentence written by a human being, or an artificial intelligence?

"AI will revolutionize people's lives and many industries by streamlining processes, optimizing decision-making, and fostering unprecedented innovation; however, we must remain vigilant against the potential risks of job displacement, ethical dilemmas, and unchecked autonomy."

Even a couple of years ago, just asking this question might have seemed absurd—but the launch of artificially intelligent chatbot ChatGPT in November 2022 sparked a worldwide frenzy around "generative AI", where artificial intelligence generates content such as text. In this case, the above sentence was written by ChatGPT, in response to a simple text prompt about the future of AI. ChatGPT (GPT stands for Generative Pre-trained Transformer) is built to offer human-like responses to questions, writing everything from reports to song lyrics. ChatGPT is a web interface for the large language models GPT-3 and GPT-4. These use a neural network which mimics the structure of the human brain, trained on huge amounts of real-world text. It generates text by predicting what the next word is likely to be. ChatGPT can also write computer code. In one demonstration, GPT-4 was shown a scanned scribbled sheet of instructions and was able to turn that into a working website.

A Global Phenomenon

The excitement around generative AI has been widely compared to the start of the internet boom in the early nineties. In the very near future, AI advocates

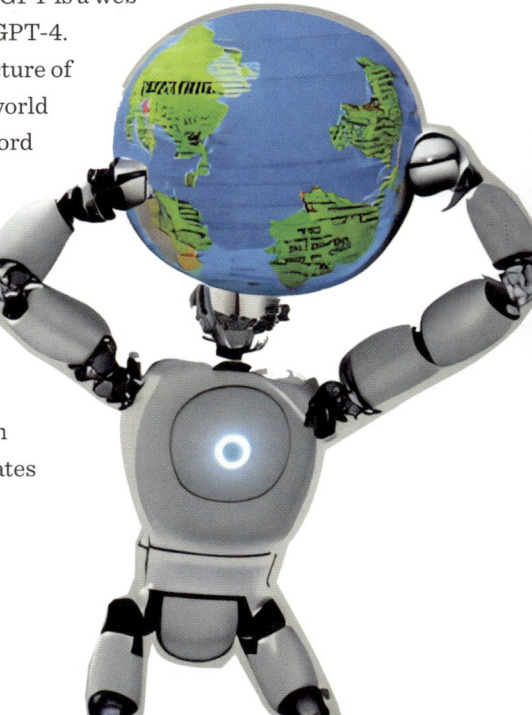

predict, such bots could reshape the way we search the internet, the way we interact and the way internet content is generated. Just weeks after launch, ChatGPT averaged 13 million users per day, making it the fastest-growing internet app of all time. ChatGPT's unearthly abilities have rapidly shaken up old certainties within education and information. ChatGPT can write essays in seconds, and the GPT-4 update is able to pass exams to qualify for the American bar with 90% scores, plus dozens of other university-level qualifications. Within weeks of its release, there were entire novels written by ChatGPT available on Amazon, and one science fiction publisher had to suspend submissions as it was bombarded with short stories written by AI, after an online influencer advertised it as a get-rich-quick scheme.

ChatGPT was not the only AI model to cause a stir. 2022 also saw the launch of art "bots" such as Craiyon, Stable Diffusion and OpenAI's DALL-E, which use huge libraries of images to generate paintings and photographs. In fact, the illustration on page 169 was generated by Craiyon. These models are already controversial, however: several large-scale lawsuits aim to challenge the legality of using human-created art to "feed" such AI tools.

OpenAI was founded in 2015 as a non-profit organization, with a billion dollars pledged by investors including Peter Thiel and Elon Musk. However, it "transitioned" to for-profit in 2019 to attract more investment (and grant employees stakes in the company) despite having previously said that its non-profit stance would allow the company to have a "positive human impact". OpenAI secured $10 billion in investment from Microsoft in 2023. Google and Facebook's parent company Meta unveiled their own rivals to ChatGPT—Bard and Llama.

New Possibilities, New Risks

OpenAI's Sam Altman described generative AI as "the greatest force for economic empowerment" that the world has ever seen. But Altman and others have warned that such models could replace large numbers of white-collar jobs, with lawyers, for example, already using large language models to generate summaries and draft documents, and various journalism organizations experimenting with AI-generated text.

But the technology potentially carries even more serious risks. With bots such as ChatGPT able to conduct fluent conversations in flawless English, experts have said that the technology could herald a new era of fraud and misinformation, where it's impossible to tell if anyone is who they say they are online.

The nature of these tools—that they produce answers by "guessing" the likeliest response, without understanding, leads to other problems. The tools need to be carefully trained and equipped with "guard rails" otherwise they are prone to giving alarming advice—such as how to self-harm, or buy unregistered guns. It seems that the problems that plagued earlier "bots" such as Microsoft's Tay (see p158) haven't entirely gone away.

The models also lie—a phenomenon known as "AI hallucination" —where AIs, which aim to generate a fluent and convincing answer, make up facts in order to do so. One of Google's demos of its Bard software saw the bot make up facts about NASA's James Webb space telescope—sending Google's share price plunging. But the technology is evolving rapidly.

Is the dream of a human-like artificial intelligence about to come true? OpenAI claims that next year, the new GPT-5 may be able to pass the Turing test for true artificial intelligence (see p85).

Altman has said that Open AI is working towards an AGI (Artificial General Intelligence), capable of human-like thought. Their mission statement is "to ensure that artificial general intelligence—AI systems that are generally smarter than humans—benefits all of humanity." The company also warns of a "serious risk of misuse, drastic accidents and societal disruption" but says, "we do not believe it is possible or desirable for society to stop its development forever."

Given the current rate of AI development, the possibility of stopping appears remote: the genie is very much out of the bottle.

Index

Adler, John, 104–6
agricultural automation, 34–6
"AI winter," 90
Altman, Sam, 169–71
Ancient Greece, 12–13
Angle, Colin, 136–8
Antikythera, 17–19
Aristotle, 12–13
art bots, 169–71
artificial intelligence, 8, 54–7,
　80–82, 83–5, 89–91, 100,
　119–21, 158–9, 163–5, 169–71
Asimov, Isaac, 58, 74–6, 95, 97
astronomical calculation, 17–19
automata, 12–13, 14–16, 20–22,
　26–8, 40–2, 54–7
automating thought, 23–5
automation, 46–8
autonomous robots, 133–5

Babbage, Charles, 40, 43–5
Badger, Julia, 150–1
Bayes, Thomas, 37–9
Bayes's Theorem, 37–9
behavior of robots, 74–6, 107–9,
　163–5
Berkeley, Edmund, 80–82
bi-pedal robots, 127–9
Breazeal, Cynthia, 110–12
Brooks, Rodney, 107–8, 136–8
Bualat, Maria, 166–8

Campbell, Murray, 119–21
Čapek Karel, 58–9, 75
cars, self-driving, 60–2, 142–4
Charles, Joseph Marie, 40–2
chatbots, 158–9, 169–71

ChatGPT, 169–71
chess, 54–7, 119–21, 155, 157
Chubbuck, John, 92–4
citizenship, of robots, 160–2
clockwork puppets, 29–31
computers, playing games,
　119–21, 155–7
computers: AlphaGo, 9, 155–7
　Deep Blue, 91, 116, 119–121,
　155
　DeepMind, 88, 155–7
　El Ajedrecista, 54–7
　ELIZA, 84
　ENIAC, 77–9
　Goostman, Eugine, 84–5
　MANIAC, 79
　Shakey, 73
　Simon, 81–2
　SNARC, 86–8
computing, 43–5
Craiyon, 170
Cyberdyne, 145–7
CyberKnife, 104–6

Da Vinci, Leonardo, 26–8
DALL-E, 170
DARPA Grand Challenge, 142–4
Devol, George, 95–7
digital computing, 77–9
Doi, Toshitada, 124–6
domestic robots, 107, 136–8,
　169–71
drones, radio-controlled, 49–51
drones: MQ-9 Reaper, 130–2

Engelberger, Joseph, 95, 97
exoskeleton, 144–7

fiction, about technology 12, 58,
　66–8, 86, 110, 127, 145, 166
Fujita, Masahiro, 124–6

Gamma Knife, 105–6
Gannon, Madeleine, 163–5
General Atomics, 130–2
Generative AI, 169–71
Giant Brains or Machines That
　Think, 80–82
Go (game), 155–7
Greiner, Helen, 136–8
grief, over robots, 124, 139–41

HAL, 144–7
Hanson, David, 160–2
Hassabis, Demis, 155–7
Hero of Alexandria, 14–16
Hoe, Richard March, 46–8
Holberton, Frances, 77–9
Holland, Owen, 133–5
Houdina, Francis P., 60–2
Hsu, Feng-hsiung, 119–21
humanoid robots, 63–5, 160–2

Jacquard loom, 13, 40–42
Jacquard, see Charles, Joseph
　Marie

karakuri dolls, 29–31
Kasparov, Garry, 91, 116, 119–121,
　155
Kelly, Ian, 133–5
Kitano, Hiroaki, 116–8
Kubrick, Stanley, 86, 145

Lang, Fritz, 66–8
law enforcement, by robots, 152–4
Laws of Robotics, 74–6, 95
Lee, Peter, 158–9
Leibniz, Gottfried Wilhelm, 24–5
lifesaving robots, 104–6
Lightning Press, 46–8
Llul, Ramon, 9, 23–5
Lovelace, Ada, 43–5

Mars rover, 139–41
Matarić, Maja, 107–9
Mauchly, John, 77–9
McCarthy, John, 89–91
Melhuish, Chris, 133–5
Metropolis, 66–8
military robots, 130–2, 146
Minsky, Marvin, 86–8, 89
ibn Musa ibn Shakir, Ahmad, al-Hasan, and Jafar Muhammad, 20–22

navigating robots, 100–103
neural computing, 86–8

Omi, Takeda, 29–31

Pollard, Willard, 69–71
privacy concerns, 153–4
probability, 37–9
Programmed Article Transfer device, 95–7
propulsion, 113–15

R.U.R. (Rossum's Universal Robots), 58–9, 75

radiosurgery, 104–6
RoboCup, 116–8
Robonaut 2, 150–1
robot "learning," 92–4, 107–9, 155–7
robot arms, 69–71, 95–7
robot arms: GammaKnife, 104–6
the Unimate, 95–6
robot animals, 113–5, 124–6
robotics, general 58–9
Robots novels, 95, 74–6, 95, 150
robots, the threat of, 76, 130–2
robots: Aibo, 116, 117, 124–6
Asimo 127–9
Astrobees, 9, 166–8
the Beast, 92–4, 127
Digidog, 154
ElliQ, 169–71
Ferdinand, 92–4
Kismet, 110–12
Kiva, 118
Manus, 165
Mars rover Opportunity, 139–41
Mimus, 163–5
Robotuna, 113–15
Roomba, 107, 136–8
Shakey, 100–103, 127
SlugBot, 113–5
Sophia, 160–2
Televox, 63–5
Toto, 107–9, 137
Rosen, Charles, 100–103
rotary printing press, 46–8

Sankai, Yoshiyuki, 29, 144–7

science fiction, 58–9, 66–8, 74–6, 145, 152
security robots, 152–4
Sedol, Lee, 155–7
seed drill, 34–6
Shigemi, Satoshi, 127–9
social robots, 110–12
Sorayama, Hajime, 67, 126
space robots, 8, 139–41, 150–1, 166–8
Squyres, Steve, 139–41
Stable Diffusion, 169–71
steam power, 16, 46–8
Stephens, Stacy, 152–4

teleautomaton, 49–51
Tesla, Nikola, 49–51
The Dartmouth Conference, 89–91
Thrun, Sebastian, 142–4
Torres-Quevedo, Leonardo, 54–7
Triantafyllou, Michael, 113–15
Tull, Jethro, 34–6
Turing Test, 83–5
Turing, Alan, 9, 83–5

Vaucanson, Jacques, 28, 41
volvelle, 23–5

walking aids, 144–7
walking robots, 127–9
Wensley, Roy J., 63–5

Glossary

Algorithm—A set of instructions to be followed in computer operations

Analytics—The discovery of meaningful patterns in data

Artificial intelligence—Intelligence demonstrated by machines, rather than the natural intelligence of humans

Artificial general intelligence—An AI capable of learning or understanding anything that a human being can

Automaton—A mechanical device made to resemble a human being

Autonomous vehicle—A "self-driving" vehicle which can travel without human input

Chatbot—Software made to chat in online settings in imitation of a human being

Degrees of freedom—The amount to which a robot (or robot arm) can move itself in different dimensions

Effector—A device or tool attached to a robot's limb to enable it to complete a task

Gynoid—A robot designed to look like a human woman

Industrial robot—A preprogrammed robot "arm" designed to move parts, tools or materials

Layered control system—A control system where complex controls are "layered" on top of a simple control system

Manipulator—A robot "hand" that can grip or pick up objects

Karakuri doll—Dolls made in Japan that use clockwork to perform humanlike actions such as drinking tea

Machine learning—Computer systems that are able to "learn" and adapt without following specific instructions

Natural language—Where software understands (or talks in) normal, spoken language rather than commands

Neural network—Computer networks loosely modeled on the structure of the human brain

Swarm robotics—Large numbers of small, simple robots that work together

Three Laws of Robotics—Rules to prevent robots harming their human masters, created by science fiction writer Isaac Asimov

Turing Test—A logical test used to determine whether someone you are talking to is a robot or a human, which was proposed by British scientist Alan Turing

Sources

Chapter 1

Aristotle, *Politics* (Translated by Benjamin Jowett) (Oxford, Oxford University Press, 1920)

Homer, *The Iliad* (Translated by Barbara Graziosi) (Oxford, Oxford University Press, 2011)

Hero of Alexandria, *Pneumatics* (Translated by Bennet Woodcroft) (London, Charles Whittingham, 1861)

Freeth, Tony et al, "A Model of the Cosmos in the ancient Greek Antikythera Mechanism," *Scientific Reports*, 2021

Banu Musa Ibn Shakir *The Book of Ingenious Devices* (Translated by Donald R Hill) (D Reidel Publishing Company, Boston, 1979)

Karr, Suzanne *Constructions Both Sacred and Profane* (Yale University Library Gazette, 2004)

Hendry, Joy *Japan at Play* (London, Routledge, 2002)

Chapter 2

Tull, Jethro, *Horse-hoeing Husbandry Or, An Essay on the Principles of Vegetation and Tillage. Designed to Introduce a New Method of Culture* (A Millar, 2007)

Bayes, Thomas, "Essay Toward Solving a Problem in the Doctrine of Chances" (Royal Society, 1763)

De Fortis, Francois-Marie, "Eloge Historique de Jacquard" (Creative Media Partners, 2018)

Meabrea, Luigi Frederico, Lovelace, Ada, "Sketch of the Analytical Engine invented by Charles Babbage ... with notes by the translator" (1843, digitized 2016)

Hoe, Robert, *A Short History of the Printing Press* (Wentworth Press, 2021)

Tesla, Nikola, "Method of and Apparatus for Controlling Mechanism of Moving Vessels or Vehicles," U.S. Patent US613809A

Chapter 3

Čapek, Karel, *R.U.R.* (Rossum's Universal Robots) (Translated by Claudia Novack) (London, Penguin, 2004)

The New York Times, "Houdini Subpoenaed Waiting to Broadcast; Magician Must Appear in Court on Charge That He Was Disorderly in Plaintiff's Office," July 23, 1925

Popular Science Monthly, "Machines That Think" (January 1928)

Von Harbou, Thea, *Metropolis* (New York, Dover, 2015)

The New York Times, "Brigitte Helm, 88, Cool Star of Fritz Lang's Metropolis," 1996

Pollard, Willard V., "Position Controlling Apparatus," U.S. Patent B05B13/0452

Moran, Michael, "Evolution of Robotic Arms," *Journal of Robotic Surgery* (2007)

Zuse, Konrad *The Computer: My Life* (Berlin, Springer Science & Business Media, 2013)

Leslie, David, "Isaac Asimov: centenary of the great explainer," *Nature* (2020)

Berkeley, Edmund, *Giant Brains, or Machines That Think* (New Jersey, Wiley, 1949)

Chapter 4

Koerner, Brendan, "How the World's First Computer Was Rescued From the Scrap Heap," *Wired* (2014)

Turing, Alan, "Computing Machinery and Intelligence," *Mind* (1950)

Bernstein, Jeremy, "Marvin Minsky's Vision of the Future," *The New Yorker* (1981)

McCarthy, Joseph, "A Proposal For The Dartmouth Summer Research Project on Artificial Intelligence," Dartmouth (1955)

Malone, Bob, "George Devol: A

Life Devoted to Invention, and Robots," *IEEE Spectrum* (2011)

Markoff, John, "Nils Nilsson, 86, Dies; Scientist Helped Robots Find Their Way," *The New York Times* (2019)

Chapter 5

McCutcheon, Stacey Paris, "Neurosurgeon John Adler is a reluctant entrepreneur,' Stanford News (2018)

Matarić, Maja J. The Robotics Primer (Cambridge, Massachusetts, MIT Press, 2007)

Cheshire, Tom, "How Cynthia Breazeal is teaching robots how to be human,' Wired (2011)

"A Brief History of RoboCup," RoboCup.org

Thomson, Elizabeth, "RoboTuna is first of new "genetic' line," 1994, news.mit.edu

Anderson, Mark Robert, "Twenty years on from Deep Blue vs Kasparov: how a chess match started the big data revolution," The Conversation (2017)

Chapter 6

"Sony Launches Four-Legged Entertainment Robot "AIBO" Creates a New Market for Robot-Based Entertainment," Sony Corporation (1999)

Ackerman, Evan, "Honda Halts ASIMO Development in Favor of More Useful Humanoid Robots," IEEE Spectrum (2018)

"MQ-9A "Reaper" Persistent Multi-Mission ISR," General Atomics

Rose, Gideon, "She, Robot: A Conversation with Helen Greiner," Foreign Affairs (2015)

"NASA's Record-Setting Opportunity Rover Mission on Mars Comes to End," NASA.gov

Chandler, David, "MIT finishes fourth in DARPA challenge for robotic vehicles," new.mit.edu

Thrun, Sebastian, "Stanley: The Robot that won the DARPA Grand Challenge," Journal of Field Robotics (Wiley Periodicals, New Jersey, 2006)

"What's HAL: The World's First Wearable Cyborg," Cyberdyne.jp

"Robonaut 2 Technology Suite Offers Opportunities in Vast Range of Industries," NASA.gov

Chapter 7

Design Museum, "Q and A with Madeleine Gannon," designmuseum.org

"Police Robots Are Not a Selfie Opportunity, They're a Privacy Disaster Waiting to Happen," Electronic Frontier Foundation

Metz, Cade, "What the AI Behind AlphaGo Can Teach Us About Being Human," Wired.com

"AlphaGo: the Story so Far," deepmind.com

Schwartz, Oscar, "In 2016, Microsoft's Racist Chatbot Revealed the Dangers of Online Conversation," IEEE Spectrum (2019)

Reynolds, Emily, "The agony of Sophia, the world's first robot citizen condemned to a lifeless career in marketing," Wired.com, 2018

"NASA's Astrobee Team Teleworks, Runs Robot in Space," NASA.gov

"Planning for AGI and Beyond", openai.com (2023)